Microbial Communities in Aquaculture Ecosystems

Nicolas Derome
Editor

Microbial Communities in Aquaculture Ecosystems

Improving Productivity and Sustainability

Springer

Editor
Nicolas Derome
Département de biologie (IBIS)
Université Laval
Quebec, QC, Canada

ISBN 978-3-030-16189-7 ISBN 978-3-030-16190-3 (eBook)
https://doi.org/10.1007/978-3-030-16190-3

This Springer imprint is published by the registered company Springer Nature Switzerland AG
The registered company address is: Gewerbestrasse 11, 6330 Cham, Switzerland

Preface

For the last two decades, the demand for fish proteins has dramatically increased and will certainly continue to rise along this century. However, such a rising demand for fish products implies significant ecological risks regarding capture fisheries. Furthermore, this increasing need for fish and seafood can no longer be met by capture fisheries alone as several fish stocks have been depleted: according to the FAO, almost 30% of global fish stocks are already harvested at an unsustainable level, whereas only 10% of fish stocks are still exhibiting abundances above their maximum sustainable yield. Also, because the overall capture fishery production is peaking since the late 1980s, the increase in fish protein demand translated into an ever-growing aquaculture production. Consequently, aquaculture is currently matching capture fishery landings with 47% of total fish production in 2016 (FAO 2018). Undoubtedly, aquaculture is becoming the leading mode of fish production for human consumption.

However, similarly to capture fisheries, such a tremendous increase of aquaculture production is accompanied with serious ecological risks. By operating mostly on coastlines or near inland rivers or ponds, aquaculture systems exert an impact on natural habitat biodiversity and productivity. For instance, high levels of food waste contribute to eutrophication by nitrogen and phosphorus release. In addition, the high density of farmed stocks increases disease episodes, and current methods of sanitary control, mostly relying on chemicals including antibiotics, persist in effluent water and contribute to the release of drug-resistant pathogenic strains into the wild.

Although accelerated aquaculture development raises enormous challenges regarding the impact of environmental degradation on the resource base, it is nonetheless the only option to provide fish food to a world population, which is expected to reach nine billion people in 2050. The path that the industry is taking today will therefore have far-reaching environmental implications for years to come. In the near future, aquaculture industry development will have no choice but to become truly sustainable to secure social, economic, and environmental benefits. Among innovations in various fields, new techniques and technologies were recently developed to mitigate pollution, such as integrating filter feeders like shellfish and/or nitrogen consumers like in aquaponics, biofiltering in recirculating aquaculture systems (RAS), food depleted in phosphorous, etc. Also, an increasing number of fish farmers reduce their dependency on forage fish by replacing fishmeal with plant proteins.

Aside from its impact on natural habitat biodiversity and productivity, the high density of farmed stocks increases exposure to unprecedented disease outbreaks, while current methods of sanitary control, mostly relying on chemicals including antibiotics, reach their limits in terms of both efficiency and sustainability. Therefore, to become truly sustainable, the aquaculture industry must adopt alternative strategies to control disease occurrence. This is where aquaculture microbiology enters into the scene.

The purpose of this book is to explore how recent aquaculture microbiology, especially microbiota research, paved the way to a highly integrated approach to understand complex relationships between farmed fish and their associated and environmental microbial communities at the frontier between health and disease. In this respect, the recent development of high-throughput sequencing technologies, combined with the rise of bioinformatics, has raised the curtain on an unsuspected microbial diversity, in terms of both taxonomy and functions. For instance, animal models such as the zebrafish (*Danio rerio*, Cyprinidae) revealed tight functional interactions between hosts and their microbiota such as regulating metabolism, immune system maturation, and via the vagus nerve, brain development and various behaviors. The increasing awareness that microbiota extensively contributes to host biology led the scientific community to rethink the study of any organism in a much more integrated way. The most recent advances in animal microbiota studies, including fish, revealed that host–microbiota beneficial interactions can be broken by multiple stressors (including unbalanced nutrition, xenobiotics, disinfection), allowing opportunistic microbial strains to induce negative effects on the host, starting with tissue inflammation and, in turn, disease. Therefore, due to the limits of the currently available curative tools such as antibiotics, one promising pathway to develop a more sustainable aquaculture is to monitor and improve both water quality and fish health by harnessing free-living and host-associated microorganisms. The current research on fish microbiota in various aquaculture contexts, from soft to salt waters, and organisms such as finfish, shrimps, and mollusks are discussed in this book. Inherent challenges and perspectives associated with each organism are thoroughly addressed.

By tightly disentangling biotic and abiotic factors that influence the host–microbiota interactions involved in host development and physiological performance, the outcome of this emerging research field will be of benefit not only to the sustainability of the aquaculture industry but also to the safety, the traceability, and the ethical acceptance of fish products.

Quebec, QC, Canada Nicolas Derome

Contents

The Rise and Fall of Antibiotics in Aquaculture

Antony T. Vincent, Jeff Gauthier, Nicolas Derome, and Steve J. Charette

Abstract It is well known that antibiotics are playing leading roles in several areas such as human and animal health and farm production. Unfortunately, the use of antimicrobial compounds as a panacea is coming to an end. More and more antibiotic-resistant bacterial strains are listed in all spheres of activity where they were used. Aquaculture, which is a key industry in providing animal protein needs to an exponentially growing human population, is undergoing the attack of pathogens that are increasingly difficult to control due to antibiotic resistance. In this chapter, we will explore, from an aquaculture perspective, the discovery and use of antibiotics, explain the major mechanisms of antibiotic resistance and the adverse consequences of using broad-spectrum antibiotics, and finally discuss briefly on alternatives to traditional antimicrobial agents. We will also give a concrete example with the bacterium *Aeromonas salmonicida* subsp. *salmonicida*, which is a worrisome pathogen for the aquaculture industry and for which crucial discoveries have been recently made thanks to the advances in sequencing technologies.

A. T. Vincent (✉)
INRS-Institut Armand-Frappier, Bacterial Symbionts Evolution, Laval City, QC, Canada
e-mail: antony.vincent@iaf.inrs.ca

J. Gauthier · N. Derome
Institut de Biologie Intégrative et des Systèmes (IBIS), Université Laval, Quebec City, QC, Canada

Département de biologie, Faculté des sciences et de génie, Université Laval, Quebec City, QC, Canada

S. J. Charette
Institut de Biologie Intégrative et des Systèmes (IBIS), Université Laval, Quebec City, QC, Canada

Département de biochimie, de microbiologie et de bio-informatique, Faculté des sciences et de génie, Université Laval, Quebec City, QC, Canada

Centre de recherche de l'Institut universitaire de cardiologie et de pneumologie de Québec (IUCPQ), Quebec City, QC, Canada

N. Derome (ed.), *Microbial Communities in Aquaculture Ecosystems*,
https://doi.org/10.1007/978-3-030-16190-3_1

1 A Brief History of Antibiotics

1.1 Discovery of Antibiotics

The term "antibiotic" was introduced in the scientific literature by Selman A. Waksman to describe "[. . .] a chemical substance, produced by micro-organisms, which has the capacity to inhibit the growth of and even to destroy bacteria and other micro-organisms" (Waksman 1947). Although the use of antibiotics may predate actual knowledge of their existence several hundred years ago (Aminov 2010), it is generally accepted that the beginning of the antibiotic era, as we know it, corresponds to the discovery of penicillin (produced by the ascomycete *Penicillium*) by Sir Alexander Fleming in 1929 (Fleming 1929). The Second World War was a major conflict that motivated scientific research and technological advances, including the antibiotic research field (Quinn 2013). At the end of the Second World War, penicillin was made available for public use on a large scale and sparked a moment of great excitement where several new antibiotic compounds have been discovered and various synthetic processes developed (Nicolaou and Rigol 2018). Undoubtedly, antibiotics have played a leading role in modern medicine, and it is widely accepted that their use has saved countless lives and increased life expectancy.

1.2 First Steps in the Use of Antibiotics in Aquaculture

Aquaculture is an increasing worldwide industry that accounted for 44.1% of the total fish production in 2014 (FAO 2016). Fish and seafood farmers therefore resort on intensive farming conditions to sustain this increasing demand for fish protein and polyunsaturated fatty acids. By its very nature, intensive fish farming exposes organisms to various stressors such as high stock density and poor water quality, thus creating a favorable environment for infection by pathogens (Sundberg et al. 2016). Obviously, fighting against fish diseases is crucial to secure production.

Shortly after their introduction for human medicine, new antibiotic compounds were made available for aquaculture (Austin and Austin 2016). One of the first well-documented usages of antibiotics in aquaculture was to treat sick brook trout (*Salvelinus fontinalis*) from furunculosis, a disease caused by *Aeromonas salmonicida* (previously named *Bacterium salmonicida*), using sulfonamides (Gutsell 1946). A major breakthrough was the discovery that combinations of sulfonamides and diaminopyrimidines (sometimes designated potentiated sulfonamides) can have a synergic effect since they both inhibit different steps of the folic acid pathway (Campbell 1999). This combination of antibiotics was found to be very effective against major fish pathogens, including *Aeromonas caviae*, *A. salmonicida*, *Vibrio anguillarum*, and *Yersinia ruckeri* (McCarthy et al. 1974; Horsberg et al. 1997).

2 The Phenomenon of Antibiotic Resistance

Despite the undeniable fact that antibiotics have helped shape modern medicine, for both humans and animals, the fight against pathogenic bacteria is far from behind us (Holmes et al. 2016). In this sense, bacteria can use various mechanisms of protection against antibiotic molecules.

2.1 Why Does a Bacterium Become Resistant?

Bacteria living in communities in the same ecological niche compete for resources. It is known that certain bacteria will themselves produce antimicrobial compounds to increase their competitiveness (Hibbing et al. 2010). Consequently, it is expected that bacterial strains resistant to antibiotics are naturally present in the environment (Martinez 2009). However, a balance will exist between resistant and sensitive strains. The overuse of antibiotics in the human (Goossens et al. 2005), veterinary (Wayne et al. 2011), and food (livestock, fish, and crop farming) contexts has created a major problem by disrupting this equilibrium (Cabello 2006; Martin et al. 2015). Although some countries have banned the use of antibiotics as growth promoters, including the European Union in 2006 (Martin et al. 2015), it is estimated that 80% of the antibiotics in the United States are used for agriculture and aquaculture, often to stimulate livestock growth or administrated as prophylactic treatments (Hollis and Ahmed 2013). Although it is difficult to clearly define the sector that has contributed the most to the amplification of the antibiotic resistance phenomenon (Chang et al. 2015), this overuse has—and continues to—generate selective pressure for bacterial cells resistant to these molecules, hence making them become dominant (Fig. 1).

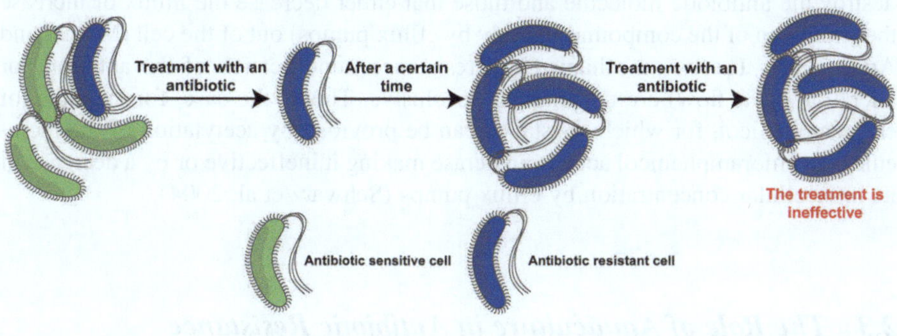

Fig. 1 Consequences of antibiotic overuse

2.2 How Does a Bacterium Become Resistant?

Bacteria develop resistance to antibiotics through two main strategies (Munita and Arias 2016). The first one is by mutation of key genes (usually those whose products are targeted by the antibiotic). A well-known example of resistance produced by mutation of a gene is against rifampicin. This antibiotic binds to the β-subunit of the RNA polymerase (encoded by the *rpoB* gene), thus blocking transcription of bacterial DNA (Campbell et al. 2001). Some non-synonymous mutations in the *rpoB* gene can decrease the affinity of rifampicin to its target, resulting in resistance (Floss and Yu 2005).

The second strategy is by acquisition of resistance genes through horizontal transfers. Compared to gene mutations (first strategy) that occur in one generation and are transmitted vertically to offspring, horizontal transfers involve the acquisition of exogenous DNA. Bacteria exchange genetic material through three main mechanisms (Fig. 2): transformation (incorporation of environmental DNA), transduction (transfer by a phage), and conjugation (contact between cells) (Holmes et al. 2016).

Although some cases of acquisition of antibiotic resistance genes by transformation or transduction have been documented, conjugation is the mechanism that contributes most to the spread of these genes (von Wintersdorff et al. 2016). In general, conjugation uses plasmids, defined as self-replicating extrachromosomal genetic elements (Actis et al. 1999), as vectors to promote the flow of genes. In fact, the biological functions of plasmids are extremely diverse and can help improve the fitness of cells by inducing, among others, antimicrobial resistance, increasing metabolic capacity, and providing virulence factors (Srivastava 2013). Finally, it is also possible to acquire antibiotic resistance genes with integrons (Fig. 2), which are site-specific recombination systems capable of recruiting genes (especially antibiotic resistance) (Deng et al. 2015).

Genes causing antibiotic resistance are divided into two groups: those that alter or destroy the antibiotic molecule and those that either decrease the influx or increase the expulsion of the compound (mainly by efflux pumps) out of the cell (Munita and Arias 2016). The two mechanisms of resistance (modification of the antibiotic or alteration of its flow) are not mutually exclusive. This is the case, for example, of chloramphenicol, for which resistance can be provided by acetylation of the molecule by a chloramphenicol acetyltransferase making it ineffective or by a decrease of its intracellular concentration by efflux pumps (Schwarz et al. 2004).

2.3 The Role of Aquaculture in Antibiotic Resistance

Aquaculture plays a central role in sustaining the demand for protein by the increasing human population (Diana 2009). To maintain the pace and remain economically profitable, fish farms need to implement intensive production

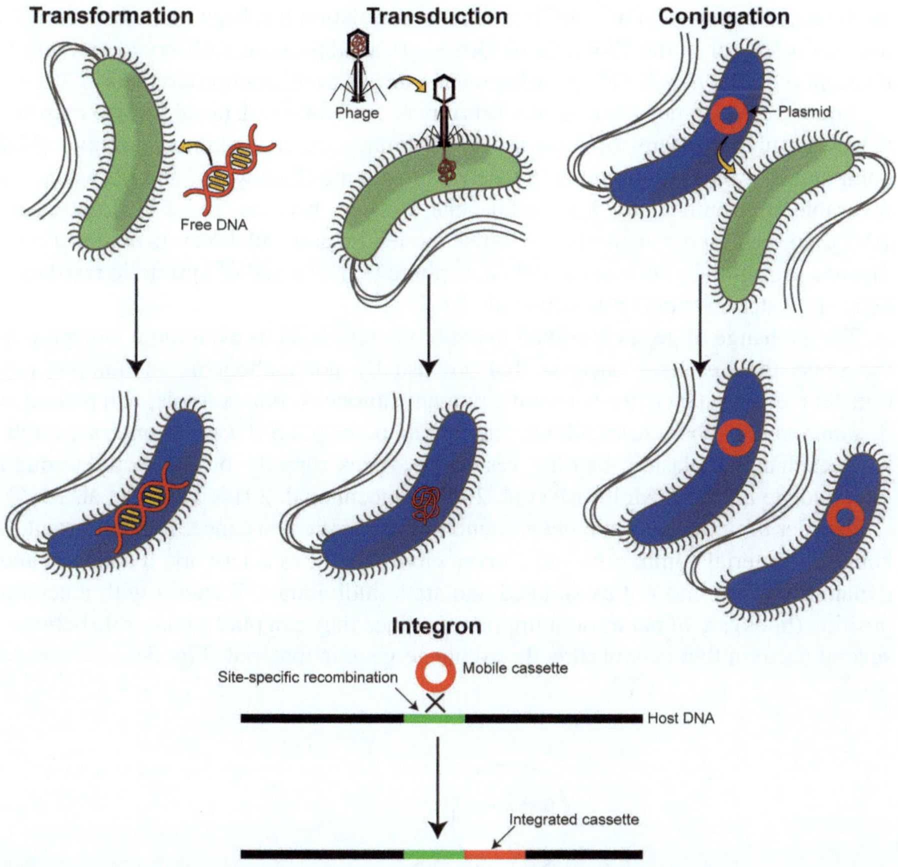

Fig. 2 Schematization of the main mechanisms used by bacteria to get or exchange DNA horizontally

conditions. This type of aquaculture, however, creates a conducive environment for the spread of disease and also promotes the emergence of pathogenic bacteria (Pulkkinen et al. 2010; Sundberg et al. 2016). As a consequence, antimicrobial agents are often used in a prophylactic manner in intensive fish farming (Cabello et al. 2013). In major producing countries like Chile, the amount of antibiotics authorized and administrated in a veterinary context can surpass the amount used in human medicine (Cabello et al. 2013).

Although this way of using antibiotics may seem attractive in the short term, the effects can be disastrous in the long term. As stated earlier, the use of an antibiotic promotes selection for resistant strains impossible to treat with the same compound. It is estimated that 80% of the administrated drugs (usually either as additives in food or by balneation) persist in the environment as active compounds (Cabello et al. 2016). A good example is oxytetracycline, which was found to persist in sediment after administration, thus causing a significant increase in bacteria resistant to this antibiotic (Samuelsen et al. 1992). A similar correlation has been made between the use of florfenicol in the Province of Quebec (Canada) and the observed amount of resistant *A. salmonicida* subsp. *salmonicida* strains to this antibiotic (Morin 2010).

Aquatic environments are ideal reservoirs of antibiotic resistance for two reasons: (1) by its high capacity of infiltration, unsanitary water containing antimicrobial compounds or other pollutants can easily contaminate clean water, and (2) water is a favorable environment for horizontal gene transfers between bacteria (Lupo et al. 2012). To this respect, using the zebrafish model, Fu and collaborators demonstrated that aquatic animal guts significantly contribute to the spread of antibiotic resistance genes in water environments (Fu et al. 2017).

The exchange of genes involved in antibiotic resistance is even more worrying in the sense that resistant bacteria that are usually not pathogenic to humans can transfer their resistance genes to some human pathogens. For example, fish pathogen *A. salmonicida* subsp. *salmonicida* and human pathogen *Salmonella enterica* might have exchanged plasmid-bearing resistance genes directly or indirectly through intermediate bacteria (McIntosh et al. 2008; Vincent et al. 2014; Trudel et al. 2016).

In order to understand the phenomenon of antibiotic resistance, it is important to consider bacterial communities in a given environment as a network, a coherent and dynamic system, and not as isolated and static individuals. Bacteria with a central position (hubs) are of paramount importance since they can play a relay role between several bacteria that cannot directly exchange genetic material (Fig. 3).

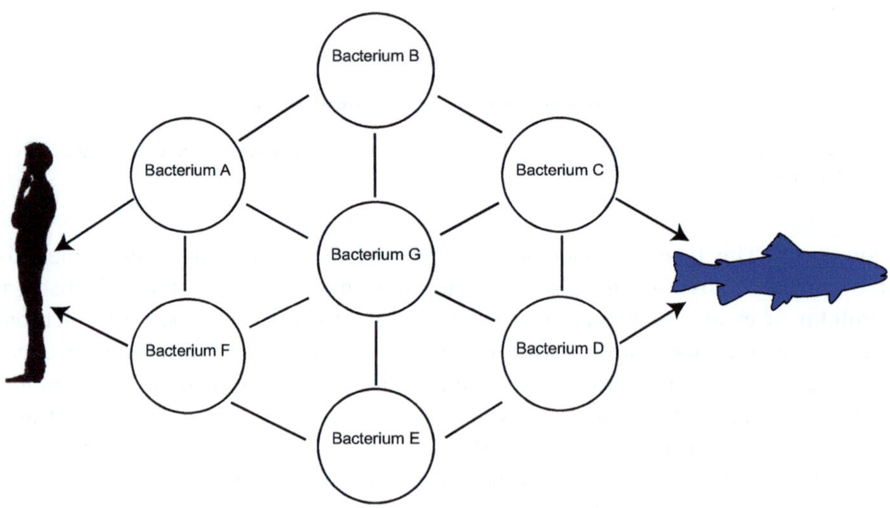

Fig. 3 Schematization of a bacterial network and how pathogens of unrelated hosts can exchange genetic material through this network

3 Adverse Effects of Antibiotherapy on Fish Microbiota

3.1 *Roles of the Microbiota in Fish Health*

Like all pluricellular organisms, fish live in close association with resident microbial communities (hereby called microbiota) composed of hundreds of microbial species. It is known that the rainbow trout microbiota is composed of 52 core bacterial lineages (Wong et al. 2013) with significant variation with respect to body site. For example, up to 199 genera can be found on rainbow trout skin (Lowrey 2014). The number of unique functional genes harbored by the microbiota surpasses the number of host genes by a 100-fold order of magnitude (Tsai and Coyle 2009). This large set of genes provided by the microbiota can complement (or even provide) metabolic pathways for nutrient metabolism (Enjalbert et al. 2017), host immunity (Belkaid and Hand 2014), and even cognitive and behavioral modulation (Carabotti et al. 2015). Those beneficial contributions of the microbiota may be disrupted following antibiotic treatment through collateral targeting of key symbionts.

3.2 *Collateral Targeting of the Microbiota by Antimicrobial Compounds*

When an infection occurs, it can be laborious and time-consuming to identify the strain that causes the disease. Consequently, antimicrobial agents having a broad spectrum (i.e., targeting a wide range of bacterial species) are usually prioritized. Although the bacterial strain that caused the infection may be correctly targeted, a wide range of other bacteria (including commensal or mutualistic symbionts) may also be affected in a collateral manner (Fig. 4). Therefore, the intensive use of antibiotics does not only promote resistance to these compounds; it also leads to deleterious side effects. A study conducted on zebrafish revealed that long-term use of legal aquaculture concentrations of oxytetracycline and sulfamethoxazole caused adverse effects on fish gut health (Zhou et al. 2018a, b) such as a decrease in goblet cell number and antioxidative enzymes and loss of intestinal microbiota diversity. Systemic effects such as decreased resistance to infection and higher oxygen consumption rate were also observed. In mosquitofish (*Gambusia affinis*), 1-week exposure to rifampicin caused a drop in viable counts from the skin microbiota (0.02% resistant) but led to >70% resistance following less than 2 days recovery after antibiotic treatment (Carlson et al. 2017). This increase was attributed to rifampicin-mediated selection for bacteria of the Comamonadaceae family.

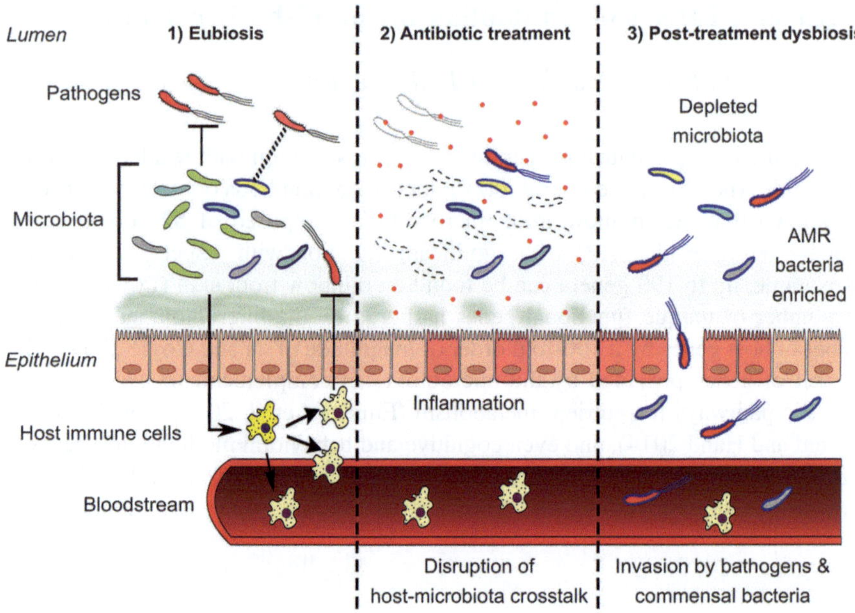

Fig. 4 Collateral targeting of bacterial symbionts by a broad-spectrum antibiotic treatment. In panel 1: Host-microbiota interactions allow mitigation of pathogens either by direct antagonism by the microbiota or by modulating the host's immune response. In panel 2: Application of a broad-spectrum antibiotic treatment targets pathogens as well as nonpathogenic symbionts from the microbiota, thereby leading to dysbiosis, i.e., a disruption of normal host-microbiota interactions. A few antimicrobial resistant (AMR) bacteria (pictured here with blue cell walls) survive, including a pathogen that acquired resistance genes from another symbiont prior to treatment (see panel 1). In panel 3: AMR bacteria thrive and occupy the gut niche previously occupied by symbionts killed by the antimicrobial treatment. As a result, the host does not adequately respond to infection

3.3 Over-elicitation of Inflammatory Responses

Symbionts from the microbiota produce natural antigens that continuously induce mucosal immune tolerance to innocuous antigens such as food proteins and molecular components of commensal bacteria (Pabst and Mowat 2012). The use of broad-spectrum antimicrobials may therefore lead to a depletion of symbionts, which help prevent active immune responses directed against the rest of the microbiota. In humans, the lack of proper immune tolerance is a contributing factor to several inflammatory diseases such as ulcerative colitis and allergy (Chistiakov et al. 2015).

3.4 Permanent Alteration of the Microbiota

The microbiota, although depending on the surrounding environment of the host, is strongly influenced by developmental and reproductive factors. In human infants, gut microbiota colonization is mainly controlled by the maternal skin microbiota, the mode of delivery, and the initial diet (i.e., colostrum and maternal milk) (Mackie et al. 1999; Fernández et al. 2013). An analogous situation was found in discus fish (*Symphysodon discus*), whose progeny feeds exclusively on maternal skin mucus in early stages of life (Sylvain and Derome 2017). In the case of human neonates, deprivation from vertical transmission routes may cause irreversible alteration of microbiota composition (Neu and Rushing 2011). In the case of discus fish, maternal mucus feeding was essential for the offspring to obtain a normal adult-like microbiota (Sylvain and Derome 2017).

Certain symbionts are only acquired during a definite part of the host's life cycle. The use of broad-spectrum antimicrobials may therefore lead to permanent changes in the host microbiota. This is exemplified by the study of Carlson and collaborators, demonstrating in *Gambusia affinis* lasting effects on mucosal microbiomes following antibiotic exposure, including the persistence of drug-resistant organisms and the inability of those microbiomes to return to a pre-treatment state (Carlson et al. 2017).

Severe consequences may arise from such a disrupted microbiota throughout development, ranging from metabolic deficiencies to increased susceptibility toward opportunistic infections (Langdon et al. 2016). In Atlantic salmon (*Salmo salar*), farmed juveniles (parrs) have a substantially lower survival rate when introduced in nature than their wild counterparts (Saloniemi et al. 2004). Interestingly, wild and farmed parrs also encountered a tremendous mismatch regarding gut microbiota composition (Lavoie et al. 2018). Furthermore, 6 months after being released into the river, stocked parrs still had a hatchery imprinting of their microbiota (Lavoie et al. 2018). Those striking differences suggest that microbiota could be another factor that could impact survival, due to its close relationship with host physiology.

3.5 Increased Carrying Capacity for Resistant and/or Pathogenic Bacteria

The intensive use of antibiotics can increase the carrying capacity of the host for pathogenic bacteria via three mechanisms. The first one, as explained above (see Sects. 3.1 and 3.2), results from the simultaneous enrichment of resistant bacteria and depletion of sensitive ones (which leaves more "room" to be occupied by resistant bacteria). The second mechanism results from the ability of bacteria to easily exchange genetic material between cells (i.e., horizontal gene transfer). Bacteria that are not pathogenic but resistant can transfer their resistance genes to sensitive but pathogenic bacteria, thereby increasing the carrying capacity of the host for pathogenic bacteria. The third mechanism is the depletion of bacteria that enhance colonization resistance. It is a broad mechanism including (1) production of mucins and defensins which prevent adherence of pathogens to mucosal tissues (Chairatana and Nolan 2017), (2) production of bactericidal and/or bacteriostatic compounds (Buffie and Pamer 2013), and (3) competition with pathogenic bacteria for the acquisition of nutrients and cofactors (Hibbing et al. 2010).

In summary, the use of antimicrobials may mitigate an outbreak in the short term but may favor the emergence of antibiotic-resistant pathogens in the long term. Furthermore, interactions between host and microbiota may be disrupted, thereby exacerbating the nefarious effects of those resistant pathogens on fish health.

4 Aquatic Pathogens Resistant to Antibiotics: The Case of *Aeromonas salmonicida*

Although it would be impossible to make a complete and exhaustive list of the aquatic pathogens for which antibiotic-resistant strains exist, there are some key problematic bacteria for the aquaculture industry. For example, several strains from the genera *Aeromonas*, *Yersinia*, *Photobacterium*, *Edwardsiella*, and *Vibrio* were listed as resistant to antibiotics, mainly through acquisition of resistance genes mediated by mobile elements (Miller and Harbottle 2018). One of the first cases of plasmid-mediated antibiotic resistance in a fish pathogen was reported in 1971 from an *A. salmonicida* strain isolated in 1959 in the United States (Aoki et al. 1971). The strain was described as resistant to sulfathiazole and tetracycline. The same study reported that these resistance phenotypes could be transferred to a strain of *Escherichia coli*. They found 15 years later that a conjugative plasmid, pAr-32, was responsible for the observed resistance (Aoki et al. 1986). It has been inferred that pAr-32 is identical to plasmid pRA3, which is the reference for IncU plasmids in addition to being found in the human and fish pathogen *Aeromonas hydrophila* (Bradley et al. 1982). In 1983, a striking correlation between the use of antimicrobial compounds and the observed resistance in strains recovered from outbreaks was reported (Aoki et al. 1983). They also found that strains of *A. salmonicida* isolated from cultured fish were more prone to have plasmid-bearing resistance genes than those from wild fish, which were shown to be mostly sensitive to the tested antibiotics.

Since then, several plasmids were found in many strains of *A. salmonicida* (Table 1) but also in other species of the same genus (Piotrowska and Popowska 2015). It is well known that advances in DNA sequencing technologies allowed us to discover and to classify genetic elements, such as plasmids, at an unprecedented pace and at relatively low cost (Vincent et al. 2017; Orlek et al. 2017). In recent years, a myriad of plasmids has been discovered and characterized in *A. salmonicida* through sequencing, many of which cause antibiotic resistance. A major discovery was the ability of *A. salmonicida* to exchange plasmids and other mobile DNA with pathogenic bacteria, such as *Aeromonas bestiarum* (plasmid pAB5S9b) and *S. enterica* (plasmid pSN254b and a class 1 integron) (Vincent et al. 2014; Trudel et al. 2016). This is even more worrying in a veterinary context given that both pAB5S9b and pSN254b cause resistance to all antibiotics approved by the Canadian

Table 1 Known plasmids that confer antibiotic resistance in *A. salmonicida* subsp. *salmonicida*

Plasmid	Length (bp)	GC%	Resistance genes	GenBank	Note	Ref.
pAsa7	5276	53.39	*cat*	NZ_KU499859	Putatively derived from the cryptic pAsa2	Vincent et al. (2016)
pAsaXII	7700	54.61	*frmA, frmB* (formaldehyde)	MF621618	Putatively derived from the cryptic pAsa2	Attéré et al. (2017)
pAsa10	9995	60.68	*tetA*	MF621616	Fragment of Tn*1721*	Attéré et al. (2017)
pRAS3.2	11,823	59.01	*tetA*	NC_003124.1		L'Abée-Lund and Sørum (2002)
pRAS3.3	11,845	58.89	*tetA*	KJ909291		Vincent et al. (2014)
pRAS3.1	11,851	58.91	*tetA*	NC_003123.1		L'Abée-Lund and Sørum (2002)
pAB5S9b	25,540	52.27	*tetH, floR, sul2, strA, strB*	KJ909292.1	Variant of pAB5S9 (*A. bestiarum*)	Vincent et al. (2014)
pASOT3	~39,000	N/A	*tetA, aadA2*	N/A		Adams et al. (1998) and L'Abée-Lund and Sørum (2001)
pASOT	~45,000	N/A	*aadA2 or dfrIIc, tetA*	N/A		Adams et al. (1998) and L'Abée-Lund and Sørum (2001)
pASOT2	~45,000	N/A	*tetA, aadA2*	N/A		Adams et al. (1998) and L'Abée-Lund and Sørum (2001)
pRAS1	~45,000	N/A	*dfrA16, tetA, sul1, qacEΔ1*	N/A	In4-like Tn*1721*-like	Sørum et al. (2003)
pAr-32	~47,000	N/A	*aadA2, sul1, catA2, qacEΔ1*	N/A		Sørum et al. (2003)
pRAS2	~48,000	N/A	*sul2, tet 31, strA, strB*	AF262622 (partial)	Tn*5393c*	L'Abée-Lund and Sørum (2000)
pAsa8	110,577	54.87	*tetA, floR, tetG, sul1, blaPSE-1, aadA2, qacEΔ1*	KX364409.1	Tn*1721* in addition to class 1 integron	Trudel et al. (2016)
pSN254b	152,216	52.47	*tetA, floR, sul1, sul2, blaCMY, aadA, strA, strB, qacEΔ1*	KJ909290.1	Variant of pSN254 (*S. enterica*)	Vincent et al. (2014)
pAsa4c	163,022	53.42	*sul1, aadA, cat, qacEΔ1*	KT033470.1		Tanaka et al. (2016)
pAsa4	166,749	52.80	*tetA*(E), *sul1, aadA, cat, qacEΔ1*	NC_009349.1		Reith et al. (2008)
pAsa4b	181,933	52.48	*tetA*(E), *sul1, qacEΔ1*	KT033469.1		Tanaka et al. (2016)

Ministry of Health's Veterinary Drugs Directorate to treat infected fish (oxytetracycline, florfenicol [a chloramphenicol analog], sulfadimethoxine/ormetoprim, and sulfadiazine/trimethoprim).

Although several plasmids found in *A. salmonicida* confer resistance to antimicrobial agents or even provide virulence factors, there are also plasmids without any known biological function. They are consequently considered as cryptic. Commonly, *A. salmonicida* subsp. *salmonicida* has three small cryptic plasmids ranging from 5.2 to 5.6 kbp and bearing either a ColE1- or a ColE2-type replicon: pAsa1 (ColE2), pAsa2 (ColE1), and pAsa3 (ColE2). These small plasmids only bear genes involved in their replication, maintenance, and mobilization. Although their presence in *A. salmonicida* subsp. *salmonicida* isolates is known since 1983 (Toranzo et al. 1983), it was in 1989 that they were named (Belland and Trust 1989), and their DNA was completely sequenced 14 years later (Boyd et al. 2003). Until recently, there was no evidence or even a clue on why these plasmids are so highly conserved throughout the isolated *A. salmonicida* strains around the world.

The plasmid pAsa11 shares high homology with pAsa3 while bearing two additional elements: the *aopP* gene, encoding a virulence factor related to the type three secretion system, and an insertion sequence (Fehr et al. 2006; Attéré et al. 2015). This plasmid is found in the majority of *A. salmonicida* subsp. *salmonicida* isolates (Attéré et al. 2015). For a long period of time, this plasmid was the sole example of a non-cryptic plasmid putatively derived from a cryptic plasmid. The situation became clearly different recently. In 2016, a plasmid named pAsa7 and bearing a gene causing resistance to chloramphenicol was published (Vincent et al. 2016). One of the salient features of this plasmid was not only the high resistance to chloramphenicol that it provides but that pAsa7 is highly similar to pAsa2. This observation reinforced the hypothesis that small cryptic plasmids, pAsa2 in the case of pAsa7, could be free high-copy receptacles for *A. salmonicida* to acquire new genes, such as for antibiotic resistance. The work of Attéré et al. in 2017 described two new plasmids, pAsaXI and pAsaXII, putatively derived from the cryptic plasmids pAsa3 and pAsa2, respectively (Attéré et al. 2017). The fact that these plasmids harbor genes involved in virulence and resistance to formaldehyde, often used as a disinfectant in aquaculture (Leal et al. 2016), combined to the existence of pAsa11 and pAsa7 let Attéré et al. state that small cryptic plasmids could be moldable vectors allowing the strains to quickly face off harsh conditions by acquiring genes involved in new functions. Additional plasmids derived from cryptic plasmids will likely be discovered in the future.

Regulation of antibiotics usages differs between countries and generates distinct pressure on the bacterial strains. This was exemplified by a recent study reporting the geographic distribution of antibiotic resistance genes and plasmids in *A. salmonicida* subsp. *salmonicida* strains from eastern Canada (Trudel et al. 2016). This study reported that several strains had plasmids, such as pSN254b and pAB5S9b, encoding genes involved in resistance to florfenicol and tetracycline, two antibiotics widely used in Canadian aquaculture. A similar observation was made by a study that investigated the epidemiology aspects of furunculosis in Denmark (Bartkova et al. 2017). They found that the resistant strains of *A. salmonicida* subsp.

salmonicida isolated in Denmark were resistant to trimethoprim and sulfonamide, two of the few antibiotics approved to treat fish infection in this country.

5 Antibiotic Alternatives and the One Health Perspective

It is now clear that in addition to better diagnostic tools, we need new treatments to fight antibiotic-resistant bacteria (Allen et al. 2014; Reardon 2015; Czaplewski et al. 2016). A group of 24 academic and industry scientists identified ten alternatives to antibiotics with enough clinical data and independent studies to believe in their approval by 2025 (Czaplewski et al. 2016). These alternatives include, for example, the use of probiotics, phages (or products derived therefrom), and antimicrobial peptides. However, this same team indicates that a budget of one and a half billion sterling pounds is needed to test and develop these ten alternatives to antibiotics. In addition, clinical trials focus primarily on human pathogens including *Clostridium difficile*, *Pseudomonas aeruginosa*, and *Staphylococcus aureus* bacteria. Therefore, it is very difficult to predict the success of these alternatives against other pathogenic and antibiotic-resistant bacteria especially in a veterinary context.

In addition to antibiotic alternatives, experts agree that antibiotic resistance must be considered in the concept of the One Health initiative (Fig. 5). This concept states that humans, animals, and the environment must be considered as a whole (Queenan

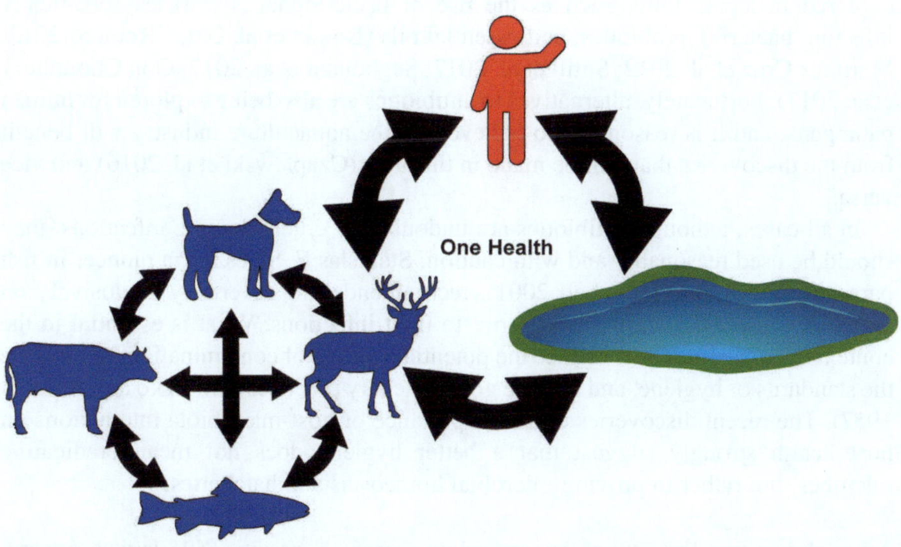

Fig. 5 Schematization of the One Health Perspective that states that animals, humans and the environment are interrelated

et al. 2016). A failure in one of these three components will inevitably affect the other two. In addition, there are some key environments that are closely related to animals and humans, as those that are aquatic (Gormaz et al. 2014). Indeed, from a "One Health" point of view, mitigating fish diseases solely through antibiotherapy without considering the rearing environment would be insufficient. Many fish pathogens such as *A. salmonicida*, *Flavobacterium* spp., and *Vibrio* spp. are opportunistic, i.e., their virulence emerges when their host's homeostasis is challenged by other abiotic and/or biotic factors (Derome et al. 2016). Therefore, one cannot truly cure an opportunistic disease without first addressing its root cause (namely, the inadequate sanitary and environmental conditions that triggered the outbreak).

6 Conclusion

Antibiotics were, and still continue to be, a crucial weapon against pathogenic bacteria. However, in addition to usually generate dysbiosis in the natural microbial flora, their introduction as a prime choice to treat bacterial infections caused the selection of strains resistant to these compounds. This is worrying in the context that several strains are now multidrug-resistant. In aquaculture, this is even more problematic since, at least for some countries, only a few antibiotics are approved as therapeutic agents.

It is increasingly important to find alternatives to antibiotics. Several options were explored in aquaculture, such as the use of bacteriophages (viruses specifically infecting bacteria), probiotics, and essential oils (Boutin et al. 2012; Romero 2012; Martínez Cruz et al. 2012; Sutili et al. 2017; Seghouani et al. 2017; Gon Choudhury et al. 2017). Fortunately, alternatives to antibiotics are also being explored for human pathogens, and it is reasonable to believe that the aquaculture industry will benefit from the discoveries that will be made in this area (Czaplewski et al. 2016) and vice versa.

In all cases, although antibiotics are undoubtedly crucial to cure infections, they should be used reasonably and with caution. Stanislas F. Snieszko, a pioneer in fish parasites and disease (Mitchell 2001), recommended to never rely exclusively on antibiotics, especially in the long term, to fight infections. What is essential in the context of aquaculture is to reduce the potential sources of contamination, to increase the standards of hygiene, and to have good hatchery practices (Snieszko and Bullock 1957). The recent discoveries on the importance of host-microbiota interactions on host health strongly suggest that a better hygiene does not mean eradicating microbes, but rather improving microbial homeostasis in hatcheries.

Acknowledgments The authors are grateful to Sammy Nyongesa (INRS-Institut Armand-Frappier, Laval, Canada) for his critical reading of the manuscript.

References

Actis LA, Tolmasky ME, Crosa JH (1999) Bacterial plasmids: replication of extrachromosomal genetic elements encoding resistance to antimicrobial compounds. Front Biosci 4:D43–D62

Adams C, Austin B, Meaden P, McIntosh D (1998) Molecular characterization of plasmid-mediated oxytetracycline resistance in *Aeromonas salmonicida*. Appl Environ Microbiol 64:4194–4201

Allen HK, Trachsel J, Looft T, Casey TA (2014) Finding alternatives to antibiotics. Ann N Y Acad Sci 1323:91–100. https://doi.org/10.1111/nyas.12468

Aminov RI (2010) A brief history of the antibiotic era: lessons learned and challenges for the future. Front Microbiol 1:134. https://doi.org/10.3389/fmicb.2010.00134

Aoki T, Egusa S, Kimura T, Watanabe T (1971) Detection of R factors in naturally occurring *Aeromonas salmonicida* strains. Appl Microbiol 22:716–717

Aoki T, Kitao T, Iemura N, Mitoma Y, Nomura T (1983) The susceptibility of *Aeromonas salmonicida* strains isolated in cultured and wild salmonids to various chemotherapeutics. Bull Jpn Soc Sci Fish 49:17–22. https://doi.org/10.2331/suisan.49.17

Aoki T, Mitoma Y, Crosa JH (1986) The characterization of a conjugative R-plasmid isolated from *Aeromonas salmonicida*. Plasmid 16:213–218. https://doi.org/10.1016/0147-619X(86)90059-4

Attéré SA, Vincent AT, Trudel MV, Chanut R, Charette SJ (2015) Diversity and homogeneity among small plasmids of *Aeromonas salmonicida subsp. salmonicida* linked with geographical origin. Front Microbiol 6:1274. https://doi.org/10.3389/fmicb.2015.01274

Attéré SA, Vincent AT, Paccaud M, Frenette M, Charette SJ (2017) The role for the small cryptic plasmids as moldable vectors for genetic innovation in *Aeromonas salmonicida subsp. salmonicida*. Front Genet 8:211. https://doi.org/10.3389/fgene.2017.00211

Austin B, Austin DA (2016) Control. In: Austin B, Austin DA (eds) Bacterial fish pathogens: disease of farmed and wild fish. Springer, Cham, pp 693–709. https://doi.org/10.1007/978-3-319-32674-0_14

Bartkova S, Leekitcharoenphon P, Aarestrup FM, Dalsgaard I (2017) Epidemiology of Danish *Aeromonas salmonicida subsp. salmonicida* in fish farms using whole genome sequencing. Front Microbiol 8:2411. https://doi.org/10.3389/fmicb.2017.02411

Belkaid Y, Hand TW (2014) Role of the microbiota in immunity and inflammation. Cell 157:121–141. https://doi.org/10.1016/j.cell.2014.03.011

Belland RJ, Trust TJ (1989) *Aeromonas salmonicida* plasmids: plasmid-directed synthesis of proteins in vitro and in *Escherichia coli* minicells. Microbiology 135:513–524. https://doi.org/10.1099/00221287-135-3-513

Boutin S, Bernatchez L, Audet C, Derome N (2012) Antagonistic effect of indigenous skin bacteria of brook charr (*Salvelinus fontinalis*) against *Flavobacterium columnare* and *F. psychrophilum*. Vet Biol 155:355–361

Boyd J et al (2003) Three small, cryptic plasmids from *Aeromonas salmonicida subsp. salmonicida* A449. Plasmid 50:131–144. https://doi.org/10.1016/S0147-619X(03)00058-1

Bradley DE, Aoki T, Kitao T, Arai T, Tschäpe H (1982) Specification of characteristics for the classification of plasmids in incompatibility group U. Plasmid 8:89–93. https://doi.org/10.1016/0147-619X(82)90045-2

Buffie CG, Pamer EG (2013) Microbiota-mediated colonization resistance against intestinal pathogens. Nat Rev Immunol 13:790–801. https://doi.org/10.1038/nri3535

Cabello FC (2006) Heavy use of prophylactic antibiotics in aquaculture: a growing problem for human and animal health and for the environment. Environ Microbiol 8:1137–1144. https://doi.org/10.1111/j.1462-2920.2006.01054.x

Cabello FC et al (2013) Antimicrobial use in aquaculture re-examined: its relevance to antimicrobial resistance and to animal and human health. Environ Microbiol 15:1917–1942. https://doi.org/10.1111/1462-2920.12134

Cabello FC, Godfrey HP, Buschmann AH, Dölz HJ (2016) Aquaculture as yet another environmental gateway to the development and globalisation of antimicrobial resistance. Lancet Infect Dis 16:e127–e133. https://doi.org/10.1016/S1473-3099(16)00100-6

Campbell KL (1999) Sulphonamides: updates on use in veterinary medicine. Vet Dermatol 10:205–215. https://doi.org/10.1046/j.1365-3164.1999.00181.x
Campbell EA et al (2001) Structural mechanism for rifampicin inhibition of bacterial rna polymerase. Cell 104:901–912. https://doi.org/10.1016/S0092-8674(01)00286-0
Carabotti M, Scirocco A, Maselli MA, Severi C (2015) The gut-brain axis: interactions between enteric microbiota, central and enteric nervous systems. Ann Gastroenterol 28:203–209. https://doi.org/10.1038/ajgsup.2012.3
Carlson JM, Leonard AB, Hyde ER, Petrosino JF, Primm TP (2017) Microbiome disruption and recovery in the fish *Gambusia affinis* following exposure to broad-spectrum antibiotic. Infect Drug Resist 10:143–154. https://doi.org/10.2147/IDR.S129055
Chairatana P, Nolan EM (2017) Defensins, lectins, mucins, and secretory immunoglobulin A: microbe-binding biomolecules that contribute to mucosal immunity in the human gut. Crit Rev Biochem Mol Biol 52:45–56. https://doi.org/10.1080/10409238.2016.1243654
Chang Q, Wang W, Regev-Yochay G, Lipsitch M, Hanage WP (2015) Antibiotics in agriculture and the risk to human health: how worried should we be? Evol Appl 8:240–247. https://doi.org/10.1111/eva.12185
Chistiakov DA, Bobryshev YV, Kozarov E, Sobenin IA, Orekhov AN (2015) Intestinal mucosal tolerance and impact of gut microbiota to mucosal tolerance. Front Microbiol 5:781. https://doi.org/10.3389/fmicb.2014.00781
Czaplewski L et al (2016) Alternatives to antibiotics-a pipeline portfolio review. Lancet Infect Dis 16:239–251. https://doi.org/10.1016/S1473-3099(15)00466-1
Deng Y et al (2015) Resistance integrons: class 1, 2 and 3 integrons. Ann Clin Microbiol Antimicrob 14:45. https://doi.org/10.1186/s12941-015-0100-6
Derome N, Gauthier J, Boutin S, Llewellyn M (2016) Bacterial opportunistic pathogens of fish. In: Hurst CJ (ed) The rasputin effect: when commensals and symbionts become parasitic. Springer, Cham, pp 81–108. https://doi.org/10.1007/978-3-319-28170-4_4
Diana JS (2009) Aquaculture production and biodiversity conservation. Bioscience 59:27–38. https://doi.org/10.1525/bio.2009.59.1.7
Enjalbert F, Combes S, Zened A, Meynadier A (2017) Rumen microbiota and dietary fat: a mutual shaping. J Appl Microbiol 123:782–797. https://doi.org/10.1111/jam.13501
FAO (2016) The state of world fisheries and aquaculture 2016. FAO, Rome
Fehr D et al (2006) AopP, a type III effector protein of *Aeromonas salmonicida*, inhibits the NF-kappaB signalling pathway. Microbiology 152:2809–2818. https://doi.org/10.1099/mic.0.28889-0
Fernández L, Langa S, Martín V, Maldonado A (2013) The human milk microbiota: origin and potential roles in health and disease. Phiarmacol Res 69:1–10. https://doi.org/10.1016/j.phrs.2012.09.001
Fleming A (1929) On the antibacterial action of cultures of a penicillium, with special reference to their use in the isolation of *B. influenzae*. Br J Exp Pathol 10:226–236. https://doi.org/10.1038/146837a0
Floss HG, Yu T-W (2005) Rifamycin—Mode of action, resistance, and biosynthesis. Chem Rev 105:621–632. https://doi.org/10.1021/cr030112j
Fu J et al (2017) Aquatic animals promote antibiotic resistance gene dissemination in water via conjugation: role of different regions within the zebra fish intestinal tract, and impact on fish intestinal microbiota. Mol Ecol 26:5318–5333. https://doi.org/10.1111/mec.14255
Gon Choudhury T, Tharabenahalli Nagaraju V, Gita S, Paria A, Parhi J (2017) Advances in bacteriophage research for bacterial disease control in aquaculture. Rev Fish Sci Aquac 25:113–125. https://doi.org/10.1080/23308249.2016.1241977
Goossens H, Ferech M, Vander Stichele R, Elseviers M (2005) Outpatient antibiotic use in Europe and association with resistance: a cross-national database study. Lancet 365:579–587. https://doi.org/10.1016/S0140-6736(05)17907-0
Gormaz JG, Fry JP, Erazo M, Love DC (2014) Public health perspectives on aquaculture. Curr Environ Health Rep 1:227–238. https://doi.org/10.1007/s40572-014-0018-8

Gutsell JS (1946) Sulfa drugs and the treatment of furunculosis in trout. Science 104:85–86

Hibbing ME, Fuqua C, Parsek MR, Peterson SB (2010) Bacterial competition: surviving and thriving in the microbial jungle. Nat Rev Microbiol 8:15–25. https://doi.org/10.1038/nrmicro2259

Hollis A, Ahmed Z (2013) Preserving antibiotics, rationally. N Engl J Med 369:2474–2476. https://doi.org/10.1056/NEJMp1311479

Holmes AH et al (2016) Understanding the mechanisms and drivers of antimicrobial resistance. Lancet 387:176–187. https://doi.org/10.1016/S0140-6736(15)00473-0

Horsberg TE, Martinsen B, Sandersen K, Zernichow L (1997) Potentiated sulfonamides: in vitro inhibitory effect and pharmacokinetic properties in Atlantic salmon in seawater. J Aquat Anim Health 9:203–210

L'Abée-Lund TM, Sørum H (2000) Functional Tn*5393*-like transposon in the R plasmid pRAS2 from the fish pathogen *Aeromonas salmonicida* subspecies *salmonicida* isolated in Norway. Appl Environ Microbiol 66:5533–5535. https://doi.org/10.1128/AEM.66.12.5533-5535.2000

L'Abée-Lund TM, Sørum H (2001) Class 1 integrons mediate antibiotic resistance in the fish pathogen *Aeromonas salmonicida* worldwide. Microb Drug Resist 7:263–272. https://doi.org/10.1089/10766290152652819

L'Abée-Lund TM, Sørum H (2002) A global non-conjugative Tet C plasmid, pRAS3, from *Aeromonas salmonicida*. Plasmid 47:172–181. https://doi.org/10.1016/S0147-619X(02)00001-X

Langdon A, Crook N, Dantas G (2016) The effects of antibiotics on the microbiome throughout development and alternative approaches for therapeutic modulation. Genome Med 8. https://doi.org/10.1186/s13073-016-0294-z

Lavoie C, Courcelle M, Redivo B, Derome N (2018) Structural and compositional mismatch between captive and wild Atlantic salmon (*Salmo salar*) parrs gut microbiota highlights the relevance of integrating molecular ecology for management and conservation methods. Evol Appl 11:1671–1685. https://doi.org/10.1111/eva.12658

Leal JF, Neves MGPMS, Santos EBH, Esteves VI. (2016) Use of formalin in intensive aquaculture: properties, application and effects on fish and water quality. Rev Aquac 0:1–15. doi: https://doi.org/10.1111/raq.12160

Lowrey LT (2014) The microbiome of rainbow trout (*Oncorhynchus mykiss*). University of New Mexico, Albuquerque

Lupo A, Coyne S, Berendonk TU (2012) Origin and evolution of antibiotic resistance: the common mechanisms of emergence and spread in water bodies. Front Microbiol 3:18. https://doi.org/10.3389/fmicb.2012.00018

Mackie RI, Sghir A, Gaskins HR (1999) Developmental microbial ecology of the neonatal gastrointestinal tract. Am J Clin Nutr 69:1035S–1045S

Martin MJ, Thottathil SE, Newman TB (2015) Antibiotics overuse in animal agriculture: a call to action for health care providers. Am J Public Health 105:2409–2410. https://doi.org/10.2105/AJPH.2015.302870

Martinez JL (2009) Environmental pollution by antibiotics and by antibiotic resistance determinants. Environ Pollut 157:2893–2902. https://doi.org/10.1016/j.envpol.2009.05.051

Martínez Cruz P, Ibáñez AL, Monroy Hermosillo OA, Ramírez Saad HC (2012) Use of probiotics in aquaculture. ISRN Microbiol 2012:1–13. https://doi.org/10.5402/2012/916845

McCarthy DH, Stevenson JP, Salsbury AW (1974) Combined in-vitro activity of trimethoprim and sulphonamides on fish-pathogenic bacteria. Aquaculture 3:87–91. https://doi.org/10.1016/0044-8486(74)90100-8

McIntosh D et al (2008) Transferable, multiple antibiotic and mercury resistance in Atlantic Canadian isolates of *Aeromonas salmonicida subsp. salmonicida* is associated with carriage of an IncA/C plasmid similar to the *Salmonella enterica* plasmid pSN254. J Antimicrob Chemother 61:1221–1228. https://doi.org/10.1093/jac/dkn123

Miller R, Harbottle H (2018) Antimicrobial drug resistance in fish pathogens. Microbiol Spectr 6:1–20. https://doi.org/10.1128/microbiolspec.ARBA-0017-2017

Mitchell AJ (2001) Finfish health in the United States (1609–1969): historical perspective, pioneering researchers and fish health workers, and annotated bibliography. Aquaculture 196:347–438. https://doi.org/10.1016/S0044-8486(01)00547-6

Morin R (2010) L'utilisation des antibiotiques pour combattre la furonculose chez l'omble de fontaine génère de l'antibiorésistance chez *Aeromonas salmonicida*. L'aquicole 15:1–6

Munita JM, Arias CA (2016) Mechanisms of antibiotic resistance. Microbiol Spectr 4:1–24. https://doi.org/10.1128/microbiolspec.VMBF-0016-2015

Neu J, Rushing J (2011) Cesarean versus vaginal delivery: long-term infant outcomes and the hygiene hypothesis. Clin Perinatol 38:321–331. https://doi.org/10.1016/j.clp.2011.03.008

Nicolaou KC, Rigol S (2018) A brief history of antibiotics and select advances in their synthesis. J Antibiot (Tokyo) 71:153–184. https://doi.org/10.1038/ja.2017.62

Orlek A et al (2017) Plasmid classification in an era of whole-genome sequencing: application in studies of antibiotic resistance epidemiology. Front Microbiol 8:182. https://doi.org/10.3389/fmicb.2017.00182

Pabst O, Mowat AM (2012) Oral tolerance to food protein. Mucosal Immunol 5:232–239. https://doi.org/10.1038/mi.2012.4

Piotrowska M, Popowska M (2015) Insight into the mobilome of *Aeromonas* strains. Front Microbiol 6:494. https://doi.org/10.3389/fmicb.2015.00494

Pulkkinen K et al (2010) Intensive fish farming and the evolution of pathogen virulence: the case of columnaris disease in Finland. Proc R Soc B Biol Sci 277:593–600. https://doi.org/10.1098/rspb.2009.1659

Queenan K, Häsler B, Rushton J (2016) A One Health approach to antimicrobial resistance surveillance: is there a business case for it? Int J Antimicrob Agents 48:422–427. https://doi.org/10.1016/j.ijantimicag.2016.06.014

Quinn R (2013) Rethinking antibiotic research and development: World War II and the penicillin collaborative. Am J Public Health 103:426–434. https://doi.org/10.2105/AJPH.2012.300693

Reardon S (2015) Antibiotic alternatives rev up bacterial arms race. Nature 521:402–403. https://doi.org/10.1038/521402a

Reith ME et al (2008) The genome of *Aeromonas salmonicida subsp. salmonicida* A449: insights into the evolution of a fish pathogen. BMC Genomics 9:427. https://doi.org/10.1186/1471-2164-9-427

Romero J (2012) Antibiotics in aquaculture – use, abuse and alternatives. In: Navarrete P, Carvalho E (eds) Health and environment in aquaculture. InTech, Rijeka, pp 159–198. https://doi.org/10.5772/28157

Saloniemi I, Jokikokko E, Kallio-Nyberg I, Jutila E, Pasanen P (2004) Survival of reared and wild Atlantic salmon smolts: size matters more in bad years. ICES J Mar Sci 61:782–787. https://doi.org/10.1016/j.icesjms.2004.03.032

Samuelsen OB, Torsvik V, Ervik A (1992) Long-range changes in oxytetracycline concentration and bacterial resistance towards oxytetracycline in a fish farm sediment after medication. Sci Total Environ 114:25–36. https://doi.org/10.1016/0048-9697(92)90411-K

Schwarz S, Kehrenberg C, Doublet B, Cloeckaert A (2004) Molecular basis of bacterial resistance to chloramphenicol and florfenicol. FEMS Microbiol Rev 28:519–542. https://doi.org/10.1016/j.femsre.2004.04.001

Seghouani H, Garcia-Rangel CE, Füller J, Gauthier J, Derome N (2017) Walleye autochthonous bacteria as promising probiotic candidates against *Flavobacterium columnare*. Front Microbiol 18:1349

Snieszko SF, Bullock GL (1957) Treatment of sulfonamide-resistant furunculosis in trout and determination of drug sensitivity. Fish Bull 57:555–564

Sørum H, L'Abée-Lund TM, Solberg A, Wold A (2003) Integron-containing IncU R plasmids pRAS1 and pAr-32 from the fish pathogen *Aeromonas salmonicida*. Antimicrob Agents Chemother 47:1285–1290. https://doi.org/10.1128/AAC.47.4.1285-1290.2003

Srivastava S (2013) Plasmids: their biology and functions. In: Genetics of bacteria. Springer, New Delhi, pp 125–151. https://doi.org/10.1007/978-81-322-1090-0_6

Sundberg L-R et al (2016) Intensive aquaculture selects for increased virulence and interference competition in bacteria. Proc Biol Sci 283:20153069. https://doi.org/10.1098/rspb.2015.3069

Sutili FJ, Gatlin DM, Heinzmann BM, Baldisserotto B (2017) Plant essential oils as fish diet additives: benefits on fish health and stability in feed. Rev Aquac 10:716–726. https://doi.org/10.1111/raq.12197

Sylvain FÉ, Derome N (2017) Vertically and horizontally transmitted microbial symbionts shape the gut microbiota ontogenesis of a skin-mucus feeding discus fish progeny. Sci Rep 7:5263. https://doi.org/10.1038/s41598-017-05662-w

Tanaka KH et al (2016) The mosaic architecture of *Aeromonas salmonicida subsp. salmonicida* pAsa4 plasmid and its consequences on antibiotic resistance. PeerJ 4:e2595. https://doi.org/10.7717/peerj.2595

Toranzo AE, Barja JL, Colwell RR, Hetrick FM (1983) Characterization of plasmids in bacterial fish pathogens. Infect Immun 39:184–192

Trudel MV et al (2016) Diversity of antibiotic-resistance genes in Canadian isolates of *Aeromonas salmonicida subsp. salmonicida*: dominance of pSN254b and discovery of pAsa8. Sci Rep 6:35617. https://doi.org/10.1038/srep35617

Tsai F, Coyle WJ (2009) The microbiome and obesity: is obesity linked to our gut flora? Curr Gastroenterol Rep 11:307–313. https://doi.org/10.1007/s11894-009-0045-z

Vincent AT et al (2014) Detection of variants of the pRAS3, pAB5S9, and pSN254 plasmids in *Aeromonas salmonicida subsp. salmonicida*: multidrug-resistance, interspecies exchanges, and plasmid reshaping. Antimicrob Agents Chemother 58:7367–7374. https://doi.org/10.1128/AAC.03730-14

Vincent AT et al (2016) Antibiotic resistance due to an unusual ColE1-type replicon plasmid in *Aeromonas salmonicida*. Microbiology 162:942–953. https://doi.org/10.1099/mic.0.000286

Vincent AT, Derome N, Boyle B, Culley AI, Charette SJ (2017) Next-generation sequencing (NGS) in the microbiological world: how to make the most of your money. J Microbiol Methods 138:60–71. https://doi.org/10.1016/j.mimet.2016.02.016

von Wintersdorff CJH et al (2016) Dissemination of antimicrobial resistance in microbial ecosystems through horizontal gene transfer. Front Microbiol 7:173. https://www.frontiersin.org/article/10.3389/fmicb.2016.00173

Waksman SA (1947) What is an antibiotic or an antibiotic substance? Mycologia 39:565–569. https://doi.org/10.2307/3755196

Wayne A, Mccarthy R, Lindenmayer J (2011) Therapeutic antibiotic use patterns in dogs: observations from a veterinary teaching hospital. J Small Anim Pract 52:310–318. https://doi.org/10.1111/j.1748-5827.2011.01072.x

Wong S et al (2013) Aquacultured rainbow trout (*Oncorhynchus mykiss*) possess a large core intestinal microbiota that is resistant to variation in diet and rearing density. Appl Environ Microbiol 79:4974–4984. https://doi.org/10.1128/AEM.00924-13

Zhou L, Limbu SM, Shen M et al (2018a) Environmental concentrations of antibiotics impair zebrafish gut health. Environ Pollut 235:245–254. https://doi.org/10.1016/j.envpol.2017.12.073

Zhou L, Limbu SM, Qiao F, Du Z-Y, Zhang M (2018b) Influence of long-term feeding antibiotics on the gut health of zebrafish. Zebrafish 15:340–348. https://doi.org/10.1089/zeb.2017.1526

Host-Microbiota Interactions and Their Importance in Promoting Growth and Resistance to Opportunistic Diseases in Salmonids

Jeff Gauthier, Camille Lavoie, Steve J. Charette, and Nicolas Derome

Abstract Salmonids are second to carps as the most important group of farmed fish, with a total annual output of over 2 million tonnes. Intensive farming practices have been developed to maximize production but at the expense of exposing farmed fish to several simultaneous stressors including frequent handling procedures, overcrowding, and poor water quality. Sanitary, prophylactic, and curative measures in an intensive farming environment are commonly used to compensate for the immune impairment that results from an over-elicited stress response. This can disrupt global interactions between the host and its microbial flora (i.e., microbiota) that play a key role in maintaining fish health in the long term. The economic importance of salmonid fish calls for a better understanding of their host-microbiota interactions to develop therapeutic tools that are less damaging for the environment and human health as well as for the fish themselves. This chapter overviews the current knowledge on factors that alter salmonid microbiomes in aquaculture and discusses the state of the art on microbial profiling and modulation, as well as current research gaps and perspectives.

J. Gauthier (✉) · C. Lavoie · N. Derome
Institut de Biologie Intégrative et des Systèmes (IBIS), Université Laval, Quebec City, QC, Canada

Département de biologie, Faculté des sciences et de génie, Université Laval, Quebec City, QC, Canada
e-mail: jeff.gauthier.1@ulaval.ca

S. J. Charette
Institut de Biologie Intégrative et des Systèmes (IBIS), Université Laval, Quebec City, QC, Canada

Département de biochimie, de microbiologie et de bio-informatique, Faculté des sciences et de génie, Université Laval, Quebec City, QC, Canada

Centre de recherche de l'Institut universitaire de cardiologie et de pneumologie de Québec (IUCPQ), Quebec City, QC, Canada

N. Derome (ed.), *Microbial Communities in Aquaculture Ecosystems*,
https://doi.org/10.1007/978-3-030-16190-3_2

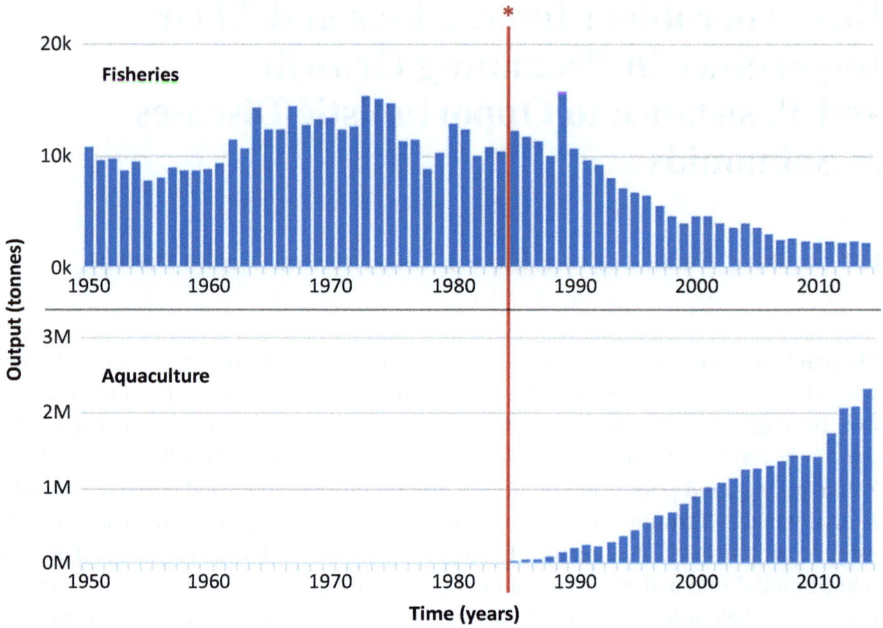

Fig. 1 Global production output for Atlantic salmon from 1950 to 2014. (asterisk) The red line indicates the tipping point (1982) from which aquaculture output started to surpass fisheries. Data from FAO Fish Stats (accessed July 4, 2018)

1 Introduction

1.1 Salmonid Aquaculture and Related Stressors

Since 2010, the global output of aquaculture reached 160 millions of tonnes, twice the amount produced by fisheries (FAO/OMS 2015). Of this number, salmonids are second to carps as the most important group of farmed fish, with a total output of over 2 million tonnes. Atlantic salmon (*Salmo salar*), which makes for two thirds of this market, has been since 1982 almost exclusively mass-produced through fish farming (Fig. 1).

This implies a strong pressure on fish farmers to keep up with high demand for this source of animal protein. Intensive farming practices have been developed to maximize production but at the expense of exposing farmed fish to several simultaneous stressors including frequent handling procedures, overcrowding, and poor water quality (Madaro et al. 2015). Unlike wild fish, captive fish cannot escape from those stressors. Even though stress is necessary to survive a danger or challenge, prolonged inescapable stress factors related to intensive rearing reduce the capacity of fish to maintain homeostasis, putting energy allocation required for reproduction, growth, and persistence on hold (Schreck 1982).

The reallocation of energy that occurs during the stress response is triggered by increased levels of the glucocorticoid hormone cortisol (Barton 2002). One of the major functions to be downregulated by elevated glucocorticoid levels is the immune system (Pickering and Pottinger 1985). Through inhibition of key transcription factors, cortisol effectively suppresses humoral factors involved in the inflammatory response and immune cell trafficking (Fast et al. 2008). Excess plasma cortisol has been shown to increase the susceptibility of brown trout (*Salmo trutta*), Atlantic salmon (*S. salar*), rainbow trout (*Oncorhynchus mykiss*), and numerous other hosts to bacterial, fungal, and parasitic diseases (Pickering and Duston 1982; Maule et al. 1989; Wilk et al. 1989; Johnson and Albright 1992).

When fish are exposed to persistent and inescapable stressors over a prolonged period of time, cortisol levels tend to remain elevated, thereby hindering the ability of stressed fish to revert back to a resting state. This was observed in Atlantic salmon, where fish exposed to handling stress had 25–75 mg/mL cortisol levels for up to 23 days post-challenge compared to near-zero levels in control groups (Madaro et al. 2015). In another study, head kidney macrophages from stressed Atlantic salmon (15 s out of water daily for 30 days) showed decreased survival when exposed to *Aeromonas salmonicida* (Fast et al. 2008).

Sanitary (e.g., egg disinfection), prophylactic (e.g., vaccination), and curative (e.g., antibiotherapy) measures in an intensive farming environment is commonly used in order to compensate for the immune impairment that results from an over-elicited stress response. When an infection occurs, it can be laborious and time-consuming to identify the strain that causes the disease. Consequently, antimicrobial agents having a broad spectrum (i.e., targeting a wide range of bacterial species) are prioritized. Although the bacterial strain that caused the infection may be correctly targeted, a wide range of other bacteria (including beneficial symbionts) are also affected. This can disrupt global interactions between the host and its microbial flora (i.e., microbiota) that play a key role in maintaining fish health in the long term.

1.2 Host-Microbiota Interactions and Their Involvement in Health

All animals live in close association with trillions of microbial cells. Their abundance is so important that they outnumber host cells by a 2:1 ratio (Sender et al. 2016). Up to 1.5% of an individual's biomass accounts for these microbes (Karlsson et al. 2013). Those constitute the host microbiota, i.e., the consortium of microbes residing on host surfaces (e.g., skin, intestines, etc.). In humans, the collective gene complement (i.e., the metagenome) of the microbiota may dwarf its host by a 150-fold factor in terms of unique functions (Qin et al. 2010). This vast gene repertoire assists the host by providing additional functions, such as metabolic pathways to digest otherwise indigestible compounds. As an example, humans cannot digest cellulose but gut bacteria in the large intestine can digest it into short-chain fatty acids (SCFAs), which humans can process with their own enzymatic toolbox (Cummings 1984). Host microbiota also contributes to immunity through (1) direct antagonism toward pathogenic microorganisms (Cherrington et al. 1991; Hammami et al. 2013), (2) signaling to the immune system (Swiatczak and Cohen 2015), and (3) reducing the carrying capacity of the host for exogenous pathogens (Kamada et al. 2013).

Benefits provided by host-microbiota interactions are highly dependent on environmental and physiological parameters. For example, acute stress responses typically shut down digestive (Mayer 2000) and immune functions (Morey et al. 2015) to react to a life-threatening danger. Alterations of these functions change both the availability of certain nutrients for microbial symbionts and reactivity of the immune system toward them. As a result, stress indirectly alters microbiota composition and, thereby, the interactions with its host.

The economic importance of salmonid fish calls for a better understanding of their host-microbiota interactions to develop therapeutic tools that are less damaging for the environment and human health (Llewellyn et al. 2014). This chapter aims, on the one hand, to present an overview of the current knowledge on the taxonomic composition (i.e., diversity and structure) of salmonid microbiota and processes governing its assembly (ontogenesis). On the other hand, the state of the art on microbial profiling and modulation will be discussed, as well as current research gaps and perspectives.

2 An Overview of Salmonid Microbiomes

A special attention has been given to the microbiota of salmonids that are significantly important in aquaculture.

*2.1 Atlantic Salmon (*Salmo salar*)*

Atlantic salmon has the most extensively characterized microbiota of all salmonids to this present day, with 17 dedicated studies published between 2007 and 2018.[1] Most of this research focused on the skin and gut microbiota, including assessments of its response to migration, nutrition, antibiotherapy, and captivity (Navarrete et al. 2008; Gajardo et al. 2017; Dehler et al. 2017; He et al. 2018).

[1]PubMed search key: ("Atlantic salmon"[Title] and "microbiota"[Title]). Last accessed: August 10, 2018

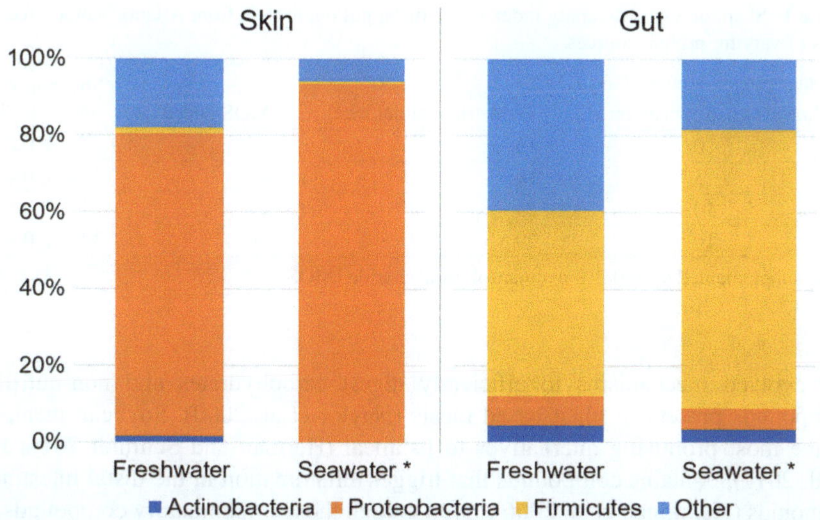

Fig. 2 Relative abundance of the dominant bacterial phyla in the Atlantic salmon skin and gut microbiota. *3 weeks post-transfer in seawater. Data from Lokesh and Kiron (2016) and Rudi et al. (2018)

2.1.1 Compositional Shifts During Freshwater-Seawater Migration

As other anadromous fish species, Atlantic salmon is exposed to two highly contrasted environments during its life cycle (i.e., freshwater and seawater), both of which differ greatly in terms of salinity, temperature, nutrient availability, and, potentially, environmental microbial exposure (Héry et al. 2014). Accordingly, major shifts in the abundance of dominant bacterial phyla were found in the skin and gut microbiota of Atlantic salmon before and after smoltification (Fig. 2). The most abundant phyla in skin and gut, respectively, *Proteobacteria* and *Firmicutes*, increase in abundance during freshwater to seawater transfers, while *Actinobacteria* decrease in both types of microbiota during this process (Lokesh and Kiron 2016; Rudi et al. 2018). However, the impact of this compositional shift on host physiology, and vice versa, remains unclear.

2.1.2 Influence of the Diet and Protein Sources

Wild Atlantic salmon feed exclusively on animal protein; juveniles start with zooplankton and feed on larger fish as they grow (Harvey et al. 2016). Accordingly, farmed salmon should be specifically fed with fishmeal as a primary source of animal protein. However, there is a growing pressure on the aquaculture industry to reduce the fishmeal content of feeds for improved sustainability and reduced cost (Rimoldi et al. 2018). Plant-based protein sources are an increasingly popular replacement for fishmeal (Newaj-Fyzul and Austin 2015). However, marine carnivorous fish have

Table 1 Shannon alpha diversity index of the distal gut microbiota from Atlantic salmon fed with diets of varying protein sources

Dietary protein sources (% $_{m/v}$)				Shannon index
Fishmeal	Soybean meal	Terrestrial animal meal	MOS added[a]	(mean ± SE)
40	0	12	+	2.33 ± 0.28
30	5	19	+	3.94 ± 0.46
18	10	29	+	2.78 ± 0.59
18	10	29	–	3.50 ± 0.45

[a]Diet supplemented with 0.2% mannan-oligosaccharide (MOS)

not evolved mechanisms to efficiently digest carbohydrates and non-nutritious compounds present in plant-based meals (Naylor et al. 2000). Soybean meal, one of the most promising alternatives to fishmeal (Herman and Schmidt 2016; Park et al. 2017), contains compounds that trigger inflammation in the distal intestine of salmonids (Heikkinen et al. 2006). Even though those inflammatory compounds can be removed by alcohol extraction, the resulting soybean protein concentrate (SPC) still alters the intestinal microbiota (Table 1). Nevertheless, fish fed with a SPC-rich diet supplemented with mannan-oligosaccharide (MOS) had an alpha diversity index more similar to fish fed exclusively with fish and terrestrial animal meals (Table 1). In another study, Atlantic salmon fed with either soybean meal, SPC, or guar meal had higher levels of lactic acid bacteria (LAB), as well as higher expression levels of proliferating cell nuclear antigen (PCNA). The cause-effect relationship between legume-based diets, LAB, and PCNA levels is still elusive, however (Gajardo et al. 2017).

2.1.3 Antibiotherapy

To our knowledge, few studies have addressed the impact of antibiotherapy on the Atlantic salmon microbiota. In 2017, a study investigated the impact of oxytetracycline (OTC), one of the most commonly used antibiotics against salmonid infectious diseases (Miranda and Zemelman 2002). OTC was administered daily in the form of medicated feed to salmon fingerlings. Microbiota composition was assessed by RFLP-PCR and sequencing of 16S rDNA amplicons. Whereas untreated microbiota was diverse and consisted mainly of *Pseudomonas*, *Acinetobacter*, *Bacillus*, *Flavobacterium*, *Psychrobacter*, *and Brevundimonas* spp., the microbiota of OTC-treated fish was dominated by *Aeromonas sobria* and *A. salmonicida.* Both species are known to harbor oxytetracycline resistance genes (Balassiano et al. 2007; Trudel et al. 2016). The latter is a well-known salmonid pathogen that causes furunculosis, a major opportunistic disease (Bullock et al. 1983). This study presented a textbook example of proliferation of opportunistic bacteria by collateral removal of competing microorganisms (see chapter "The Rise and Fall of Antibiotics in Aquaculture" for a detailed discussion on this topic). To our knowledge, no other study investigated specifically how antibiotics impact the Atlantic salmon microbiota. In this regard, there is a significant knowledge gap regarding other common antimicrobials such as florfenicol (Nordmo et al. 1998) and sulfamethoxazole-trimethoprim (Kadlec et al. 2011).

2.1.4 Captivity

Even though the bulk of salmon aquaculture is intended to produce food, government-led programs were also introduced to restore endangered salmon populations in rivers (Province of Quebec, Canada, is a notable example). Most involve stocking rivers with hatchery-reared juveniles (usually 0+ or 1+ parrs). Even though stocking parrs is preferred to stocking captive adults because of the latter's low reproductive success, captive parrs do not survive as well in the wild as their wild-born counterparts. A 2018 study revealed a substantial mismatch between the microbiota of captive (meant for stocking) and wild parrs (Lavoie et al. 2018) sampled from two different rivers. Even though community composition from wild parrs was specific to the river, captive fish (born from wild breeders from either river) were not significantly differentiated despite their distinct genetic origin. Furthermore, their microbiota composition was highly distinct from their wild fish relatives. In addition, captive parrs' microbiota was dominated by *Firmicutes* (*Lactobacillaceae*), whereas wild parr's microbiota was enriched with *Proteobacteria* (*Enterobacteriaceae*). Those results were consistent with previous studies indicating that the microbiota composition is highly associated with the diet protein source (Desai et al. 2012; Gajardo et al. 2016). As such, captive parrs are fed with commercial pellets made from vegetable proteins, a great source of carbohydrates. The latter has been associated with an increase of lactic acid bacteria (LAB) such as *Lactobacillaceae* and contributes to the divergence between captive and wild parr's microbiota composition.

As well as highlighting the substantial contribution to diet and environmental conditions on the microbiota composition, this study also confirmed that the bacterial species richness (alpha diversity) can be associated with the selective pressure of an environment. In comparison to captive parrs, wild juveniles showed a much lower diversity index and a higher homogeneity within the individual's microbiota composition, suggesting that higher selective pressure translates into a more specialized microbiota composition (Derome et al. 2006).

Interestingly, some disparities were detected when studying the network interactions of taxa according to the parr's origin. For instance, a higher proportion of negative interactions was found within captive parr's microbiota (Fig. 3). Those results are of prime interest since it has been established that negative correlations can be associated with a higher dysbiosis index (Vázquez-Baeza et al. 2016). Overall, captivity and hatchery rearing highly contribute to the microbiota composition, even for parrs from the same genetic population. Studying microbial ecology in the aquaculture field is therefore totally pertinent for assessing the effect of

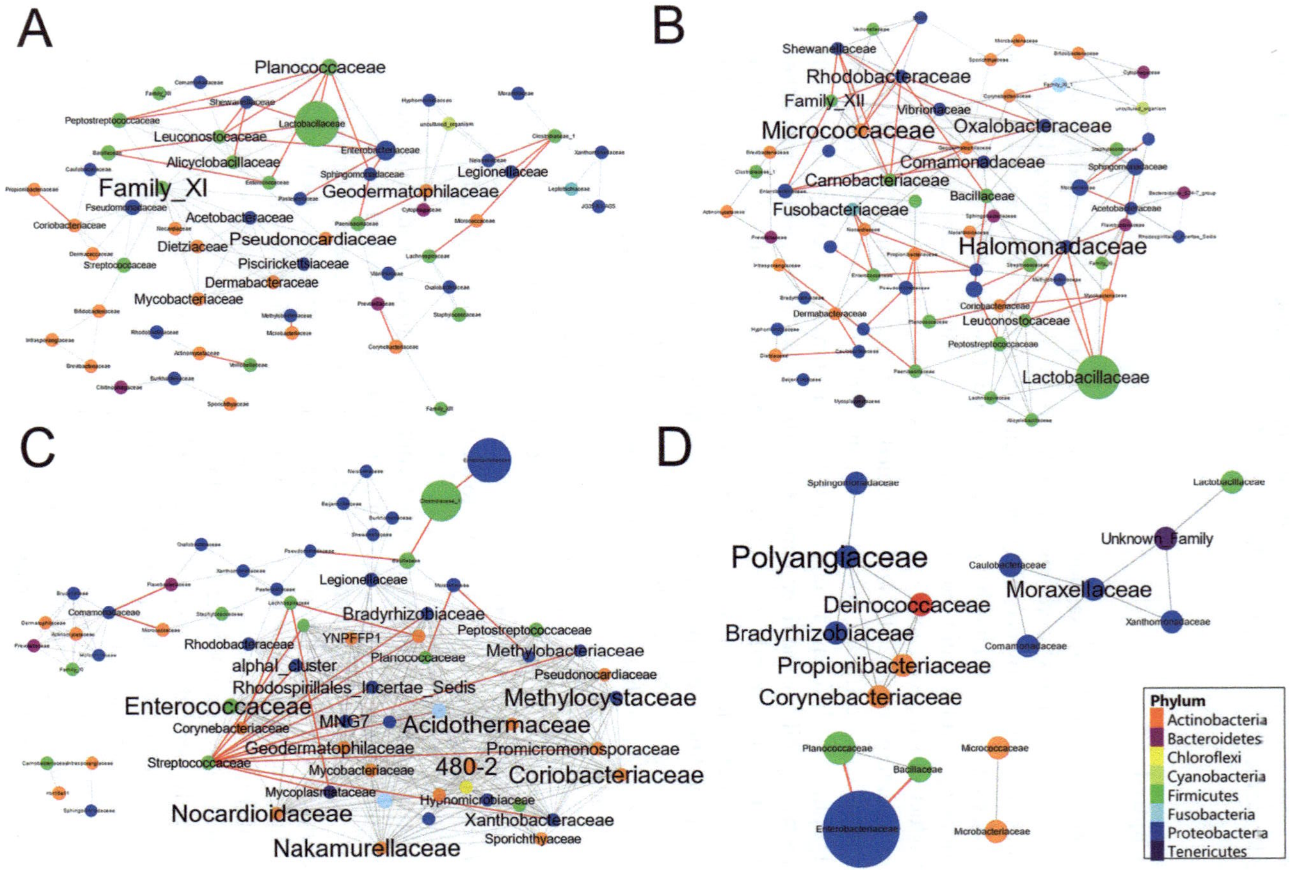

Fig. 3 Network analysis (Spearman correlation coefficient) of the microbiota at the family level depending on parrs' origin in two Canadian rivers. Node size is proportional to the relative abundance of each taxa, where low abundance taxa show more interactions than abundant taxa. Font size is proportional to the number of interactions for each taxa, illustrating keystone taxa for every environment. Negative correlations are illustrated by red edges. (**a**) Malbaie captive parrs; (**b**) Rimouski captive parrs; (**c**) Malbaie wild parrs; (**d**) Rimouski wild parrs. Data from Lavoie and Derome (2018, unpublished data)

captivity on physiology and microbiota, especially if reared fishes are meant to be released thereafter.

Since the diet of fishes appeared to be greatly associated with the host metabolism, stocking methods have been modified over time to mitigate the physiological mismatch of hatchery-reared fishes that are meant to be reintroduced in nature (Milot et al. 2013). For example, hatcheries are now raising Atlantic salmon juveniles until the alevin stage, which still have their unabsorbed yolk sac at the moment of stocking. However, hatchery rearing seems to have left a permanent imprint on the microbiota of stocked alevins, despite them being no longer exposed to an artificial environment and not being fed during the rearing (Lavoie and Derome 2018, unpublished data).

By analyzing the microbiota of wild and stocked juveniles that have been sampled in the river 4 months after stocking, differences between the microbiota of stocked and wild individuals are still highlighted. Overall, discrepancies are observed for the taxonomical composition of the microbiota as well as the diversity, suggesting a potential mismatch for metabolic functions. For instance, a higher diversity index is associated with stocked parrs' microbiota, indicating that the rearing conditions have a permanent effect on the structure of the microbiota. Even though the exact contribution of this mismatch on host fitness is unclear and deserves further investigation, hatchery rearing is proven to drive microbial processes at various levels. Acquiring a better understanding on how the microbiota is affected by the environment after stocking will certainly lead to the optimization of the conservation methods of endangered fish species.

2.2 *Rainbow Trout* (**Oncorhynchus mykiss**)

Rainbow trout is second to the Atlantic salmon as the most produced fish in salmonid aquaculture with 812,000 tonnes produced in 2014, which represents 35% of global Atlantic salmon production (FAO Fisheries and Aquaculture 2018). A total of 14 studies[2] on the rainbow trout microbiota were published from 2010 to 2018, most of which investigated the influence of diet on various physiological parameters. The remaining studies investigated the role of the microbiota in growth promotion and pathogen inhibition, as well as diet-immunity interactions and their impact on the microbiota. None of those studies addressed the impact of antibiotherapy on the microbiota composition and subsequent effects on fish health. Perhaps due to the relevance of this species in aquaculture, a special emphasis on the intestinal microbiota was found throughout most of the aforementioned studies.

[2]PubMed search key: ("rainbow trout"[Title] and "microbiota"[Title]). Last accessed: Dec 10, 2018

2.2.1 Influence of Nutrition

Farmed rainbow trout was found to possess a core gut microbiota of 52 bacterial lineages (Wong et al. 2013). This core gut microbiota was remarkably resilient to interindividual variation, diet, and rearing density changes, with no significant change in the abundance of bacterial classes. Nevertheless, the gut microbiota as a whole responds to diet composition.

One of the hot topics in rainbow trout aquaculture is the use of alternative (plant-based) protein sources to improve the sustainability of fish farming. However, those are quite rich in carbohydrates compared to rainbow trout's natural feed, which is very rich in protein (>40%) and poor in carbohydrates (<1%). A short hyperglucidic-hypoproteic stimulus (HHS) during early life stages was found to induce a long-term influence on the gut fungi (but not bacteria) profiles. Furthermore, it induced upregulation of glucose metabolism genes and downregulation of gluconeogenesis and amino acid catabolism genes in muscle tissue (Geurden et al. 2014). In the long term, HHS-treated fish did not differ in growth, feed intake, or efficiency of feed utilization. However, a significant effect on glucose homeostasis was observed. Up to 9 h after being fed the same commercial diet, HHS-treated fish had 1.5-fold higher glycemia than untreated fish. This hints to the possibility of nutritional programming as a way of optimizing the use of alternative plant-based feeds in fish farming.

The rainbow trout gut microbiota also responds to the inclusion of dietary additives. Supplementation with a plant essential oil mixture (MixOil) altered gut microbiota diversity indices and fillet quality metrics, but the link between those two remains unclear (Ceppa et al. 2018). The inclusion of organic acids in aquafeed was found to influence gut microbiota composition, but with unclear effects on host physiology (Jaafar et al. 2013).

2.2.2 Diet-Immunity Interactions

One of the main causes of mortality in rainbow trout aquaculture is *Yersinia ruckeri*, the causative agent of enteric redmouth disease (Tobback et al. 2007). The initial target organ for *Y. ruckeri* appears to be the gut (Méndez and Guijarro 2013). Accordingly, the administration of probiotic bacteria (via coated feed) enhanced resistance to this pathogen (Raida et al. 2003), but the mechanism of action was unclear.

In 2014, a Danish team investigated the missing link between microbiota, diet, and the immune response in fish challenged with *Y. ruckeri* (Ingerslev et al. 2014). Rainbow trout fry challenged by *Y. ruckeri* were split into two diet groups containing either (1) fishmeal + fish oil or (2) fishmeal/Pea meal (9:1) + rape seed oil. Microbiota composition was assessed by deep sequencing of the 16S rRNA gene, and immune gene expression was quantified by RT-qPCR. In summary, challenged fish fed with the marine-based diet had higher counts of *Yersinia* (as determined by either bacteriology or 16S profiling) and had also increased expression levels of interleukins 1-beta and 2. The plant-based diet may have had a prebiotic effect by favoring the presence of taxa that are protective against *Y. ruckeri* (Ingerslev et al. 2014). However, post-infection cumulative survival did not significantly differ between challenged fish fed either diet.

2.2.3 Pathogen Inhibition

The total cultivable microbiota of Chilean farmed rainbow trout harbors lactic acid bacteria (LAB) in high abundance (Araújo et al. 2015). Of those, 71% (mostly *Lactococcus lactis* isolates) possess inhibitory activity against one or more of the following pathogens: *Lactococcus garvieae*, *Streptococcus iniae*, *Yersinia ruckeri*, *Aeromonas salmonicida*, and *Vibrio campbellii* (Araújo et al. 2015). Whether those LAB isolates do possess inhibitory activity in vivo remains to be investigated.

2.3 *Brook Charr (*Salvelinus fontinalis*)*

The brook charr microbiota remains largely mischaracterized, except for the skin mucus (SM) microbiota, for which response to intensive rearing conditions and symbiont-pathogen interactions (including interindividual variations) were investigated (Boutin et al. 2012, 2013a, b, 2014). To our knowledge, no published studies have yet discussed the impact of diet or therapeutic tools on both microbiota structure and brook charr physiology.

2.3.1 Microbiota Structure and the Stress Response

The brook charr SM microbiota is dominated by the bacterial phyla *Proteobacteria*, *Actinobacteria*, and *Bacteroidetes* (Boutin et al. 2013b). Their relative abundance shifts abruptly when fish are exposed to hypoxia and high-density stress (Fig. 4). In addition to those abundance shifts, strong co-occurrence patterns were found. Some co-occurring genera associated with opportunistic diseases (*Psychrobacter*, *Steroidobacter*, *Pseudomonas*, *Acinetobacter*, *Aeromonas*) were specific to stressed and dead fish, whereas others (*Sphingomonas*, *Methylobacterium*, *Propionibacterium*, and *Thiobacter)* were abundant only in unstressed fish (Boutin et al. 2013b). Beneficial bacteria tended to decrease in a colinear manner following a stress event, thus resulting in an empty niche for opportunistic pathogens, which accordingly tended to increase as co-abundant groups. The role of SM microbiota in preventing infections in its host might be more important than previously thought. Indeed, several endogenous strains, in addition to those isolated from the gut microbiota, have shown inhibitory effects against common brook charr pathogens (Table 2).

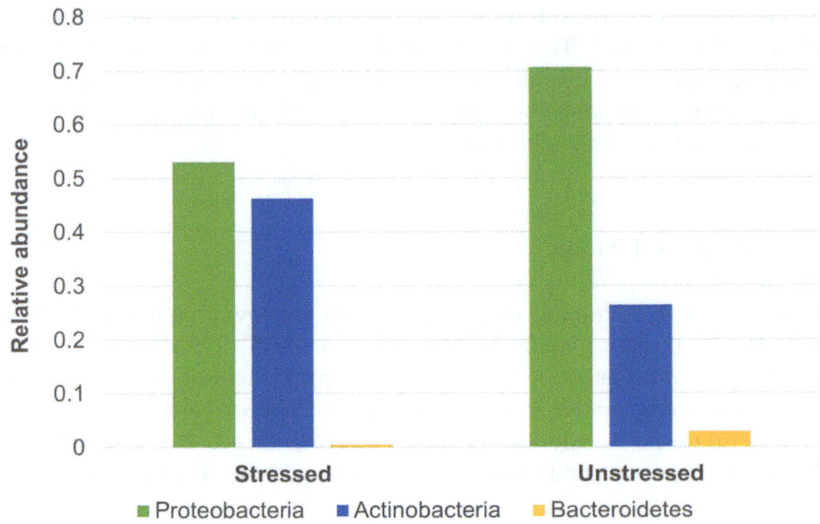

Fig. 4 Relative abundance shifts in the skin mucus microbiota of stressed versus unstressed brook charr. Stressed fish were exposed to high density (80 fishes in 10 L) until the oxygen concentration decreased to 3 mg/L (5 min). After stress exposure, fish were transferred in a new oxygenated tank to slowly recover. Data from Boutin et al. (2013b)

2.3.2 Symbiont-Based Therapeutic Tools Against Opportunistic Pathogens

An indigenous brook charr isolate from skin mucus (*Rhodococcus* sp. CPM5) decreased mortality due to pathogens *Flavobacterium psychrophilum* and *F. columnare* by 47% without disturbing the natural microbiota of skin mucus (Boutin et al. 2013a). Unexpectedly, it was not by recolonizing the skin mucus microbiota that CPM5 conferred its protective effect, but rather by colonizing the filtering mass of the recirculation system, where it may have had a positive impact on water quality, as *Flavobacterium* spp. were observed to be virtually absent from the circulating water in treated groups (Boutin et al. 2013a). In addition, CPM5-treated fish's water was dominated by *Sphingomonas* spp. unlike control tanks. Interestingly, *Sphingomonas* is the dominant bacterial genus in brook charr skin mucus. CPM5 may have indirectly improved resistance to flavobacteriosis by acting as a prebiotic for *Sphingomonas* spp., which in turn may have excluded *Flavobacterium* spp. from the surrounding water. This "symbiotic action at a distance" raised interesting questions regarding the nature of host-microbiota symbiotic relationships. Furthermore, it indicated that microbial symbionts may be recruited into novel ecological functions when readministered independently (Watson and Pollack 2001).

Other bacterial brook charr symbionts showed great promise as inhibitors of another major salmonid pathogen, *Aeromonas salmonicida* subsp. *salmonicida*

Table 2 Endogenous isolates from brook charr microbiota known to inhibit salmonid pathogens

			Known inhibitory effects			
Strain	Source	Origin	Effective against	In vitro effect?[a]	In vivo effect?[b]	References
Pseudomonas fluorescens ML11A	Skin mucus	Quebec, QC, Canada	*A. s. s.*[c]	Yes	NA	Gauthier (2016) and Gauthier et al. (2017a)
Pseudomonas fluorescens ML11B	Skin mucus	Quebec, QC, Canada	*A. s. s.*	Yes	NA	Gauthier (2016)
Pseudomonas fluorescens ML13	Skin mucus	Quebec, QC, Canada	*A. s. s.*	Yes	NA	
Aeromonas sobria TM12	Intestine	Kamouraska, QC, Canada	*A. s. s.*	Yes	NA	Gauthier et al. (2017b)
Aeromonas sobria TM18	Intestine	Kamouraska, QC, Canada	*A. s. s.*	Yes	NA	
Luteimonas sp. CP1	Skin mucus	Quebec, QC, Canada	*Flavobacterium columnare*	Yes	NA	Boutin et al. (2012)
Microbacterium sp. CP2	Skin mucus	Quebec, QC, Canada	*Flavobacterium psychrophilum*	Yes	NA	
Rhodococcus sp. CP3	Skin mucus	Quebec, QC, Canada	*Flavobacterium psychrophilum*	Yes	NA	
Microbacterium sp. CP4	Skin mucus	Quebec, QC, Canada	*Flavobacterium psychrophilum*	Yes	NA	
Rhodococcus sp. CP5	Skin mucus	Quebec, QC, Canada	*Flavobacterium columnare*	Yes	Yes	Boutin et al. (2012, 2013a)
Pseudomonas sp. CP6	Skin mucus	Quebec, QC, Canada	*Flavobacterium psychrophilum*	Yes	NA	Boutin et al. (2012)
Sphingopyxis sp. CP7	Skin mucus	Quebec, QC, Canada	*Flavobacterium psychrophilum*	Yes	NA	
Leucobacter sp. CP8	Skin mucus	Quebec, QC, Canada	*Flavobacterium columnare*	Yes	NA	
Dietzia sp. CP9	Skin mucus	Quebec, QC, Canada	*Flavobacterium psychrophilum*	Yes	NA	

[a]Evidence of in vitro inhibitory effect against pure pathogen cultures
[b]Successful decrease of mortality or morbidity when administered to challenged brook trout. *NA* not available
[c]*Aeromonas salmonicida* subsp. *salmonicida*

(*A. s. s.*). One of those, *Pseudomonas fluorescens* ML11A, was also recovered from skin mucus (Gauthier et al. 2017a) and exhibited a strong antagonistic effect across a wide range of *A. s. s.* from different geographical origins (Gauthier 2016). In addition, gut isolates belonging to the *Aeromonas sobria* (sensu stricto) species, TM12 and TM18, showed tremendous inhibitive properties against *A. s. s.* through inhibitory compound diffusion on agar (Gauthier et al. 2017b). Those two probionts were shown to increase plasma lysozyme activity, respectively, by a 1.5- to 2-fold factor in in vivo preliminary experiments without *A. s. s.* challenge, suggesting a positive impact on host innate immunity (Gauthier et al. 2016, unpublished data). However, whether the inhibitive property of those probiont strains translates into a protective effect in vivo remains to be investigated.

3 An Overview of High-Throughput Methods and Their Contribution to Microbiota Studies

High-resolution study of the microbiota has been made possible with the advent of omics-based methods (genomics, transcriptomics, proteomics, etc.) following the emergence of high-throughput DNA sequencing during the late 2000s. This technological revolution made it possible to obtain several millions of nucleotide reads from tens to hundreds of samples at once, for costs that are orders of magnitude cheaper than Sanger sequencing for an equivalent amount of data (Vincent et al. 2017a). Knowing that a major fraction of microbial diversity is unculturable, high-throughput DNA sequencing technologies proved useful to the analysis of complex microbial assemblages. Several high-throughput sequencing methods have emerged, each offering distinct elements of information on the host-microbiota complex. For this reason, an integrative research strategy should ideally use a combination of these methods.

3.1 Whole-Genome Sequencing (for Specific Microbes)

Whole-genome shotgun (WGS) sequencing aims to obtain the whole DNA sequence of a specific organism. For individual microbes isolated from the microbiota, this method requires that the organism be in pure culture in order to avoid contaminating the sequence with exogenous DNA. Consequently, the DNA of individual nonculturable microbes cannot yet be sequenced independently (see paragraph below on metagenomics).

Briefly, a pure DNA extract of the organism is sheared into fragments of a few hundred base pairs long and is then sequenced using high-throughput technology, e.g., Illumina MiSeq (Tagini and Greub 2017). Typically, this process yields several hundred thousand DNA sequences (i.e., reads) that have to be assembled de novo (i.e., without prior knowledge on source DNA) to obtain the complete sequence of the organism's genome. In most genome sequencing projects, the assembly does not reach completion and yields tens to hundreds of larger chunks (contigs) that span from a few kilobases to a few megabases in length. In fact, since 2014, there are more genomes published as incomplete drafts than complete ones (Gauthier et al. 2018) since it does not interfere much with downstream analyses.

When a draft assembly is complete, gene sequences are then annotated by similarity searches against databases of known genes. Some examples of web services offering this analysis are the NCBI Prokaryotic Genome Annotation Pipeline (Tatusova et al. 2016) and RAST (Overbeek et al. 2014). When gene annotation is complete, several downstream analyses can be applied: metabolic pathway reconstruction (Kanehisa et al. 2016), taxonomic assignment using average nucleotide identity (Konstantinidis and Tiedje 2005), phylogenomic inference (Delsuc et al. 2005), and virulence assessment (Chen et al. 2016) as well as antibiotic resistance gene prediction (Jia et al. 2017).

WGS sequencing was used, for example, to characterize the taxonomy and functions of individual bacterial symbionts in brook charr and rainbow trout which had potential as probiotic treatments against salmonid diseases (Boutin et al. 2012; Schubiger et al. 2015; Gauthier et al. 2017a, b), as well as pathogens (Reith et al. 2008; Rochat et al. 2017). However, WGS does not provide much insight on the higher organizational levels of the microbiota. The two following approaches (metabarcoding and metagenomics), respectively, address the following questions: (1) which microbes are involved and (2) how they contribute to the functional repertoire of the microbiome. Note that WGS data plays an important role in the annotation of functional data in metagenomics, as explained in the subsections below.

3.2 Metabarcoding (Who Is There?)

Metabarcoding is the massively parallel sequencing of a universal genetic marker to infer the taxonomic census of a community. It is currently the most common method in current microbiome studies, regardless of studied organisms (Garrido-Cardenas and Manzano-Agugliaro 2017; Mahato et al. 2017; Osman et al. 2018). The 16S ribosomal RNA gene is the most commonly used biomarker, due to its ubiquitous presence in bacterial, archaeal, and eukaryotic (organellar) genomes (Ju and Zhang 2015). Most, if not all, of the aforementioned salmonid-microbiota studies in Sect. 2 used 16S rRNA gene metabarcoding to assess microbiota composition.

First, DNA from the microbiota is extracted from host samples (e.g., skin mucus, gills, gut section). Then, a gene that is universally present across the widest range of organisms (a biomarker) is amplified by PCR. After this process, DNA "tags" are added to the amplified DNA products to allow "per sample" identification. PCR products are then sequenced simultaneously on a high-throughput apparatus (e.g., Illumina MiSeq or Ion Torrent PGM). There is a myriad of sample collection and DNA extraction methods that were developed in the last decade (Pollock et al. 2018), each with their own variability and biases for specific microbial groups. In a 2018 study, various combinations of gut sample collection, DNA extraction, and high-throughput sequencing were compared (Panek et al. 2018). However, optimization efforts (including the aforementioned study) have so far mostly been made for human gut microbiota studies. No methodological analysis on salmonid sample processing (skin mucus, gills, gut) has yet been published.

After sequencing, several files are produced, each containing several tens of thousands of sequence reads from a specific sample. Several computer programs are available to process the raw reads, each contributing to one or more of the steps involved in the analysis (Table 3). Typically, reads will be clustered in operational taxonomic units (OTU) under the assumption that similar sequences belong to a single microbial lineage. OTUs are then quantified by counting the reads that were involved in their construction. Then, reads are compared against a sequence database of known organisms, e.g., Greengenes (DeSantis et al. 2006), RDP (Cole et al. 2014), or ARB-SILVA (Yilmaz et al. 2014), to determine the consensus taxonomy of each OTU. Finally, an "OTU table" is obtained, on which a plethora of quantitative methods can be used to identify differences in microbiota structure between treatments, conditions, or samples (Table 3).

In summary, (16S) metabarcoding ultimately attempts to correlate the microbiota taxonomic composition to treatments dispensed to a host or several health metrics (i.e., growth rate, size, blood cell count, plasma lysozyme activity). However, it does so without providing much information on the function of microbes that are present (Zepeda Mendoza et al. 2015). Functional roles may be either (1) inferred grossly by reviewing what is known about specific families or genera in the literature or (2) inferred systematically by using "metagenome prediction" software to link taxa to whole-genome data (Langille et al. 2013; Aßhauer et al. 2015). However, most metabarcoding methods resolve taxonomy up to the genus level, and the functional repertoire of species within a single genus can be tremendously variable. For example, the genus *Pseudomonas* encompasses a wide array of mutualistic, commensal, and pathogenic species living across different habitats (soil, water, or in association with animals or plants) (Silby et al. 2011). Unsurprisingly, about only 1% of all known *Pseudomonas* genes is shared among all of its known genomes (Freschi et al. 2018). Therefore, using metagenome prediction from metabarcoding may overgeneralize the attributes of a single genus.

3.3 Metagenomics (What Are They Doing?)

Metagenomics, in principle, is not that different from WGS sequencing from a methodological point of view. However, input samples will be DNA extracts of complex communities (e.g., gut tissue sections or skin mucus swabs) instead of a single organism's DNA, leading to the untargeted sequencing of all microorganism's genomic sequences present in a given sample, in addition to the host organism's genome (Quince et al. 2017). Due to the presence of DNA molecules from tens to thousands of different taxa in one sample, several hundred thousand fragments (contigs) per sample are typically obtained. Those contigs often correspond to individual gene sequences. Those are then annotated by homology search against known sequence databases, after which post-annotation analysis and visualization methods can be applied (Table 4). This methodological approach is therefore

Table 3 Major steps in a typical 16S amplicon sequence analysis pipeline, including key software involved

Step	Sub-step	Description	Example software
Data preprocessing	Quality filtering and trimming	Removing incorrectly called nucleotides in sequence reads, as well as reads of poor quality. In a trimming approach, reads are trimmed after quality reaches a certain threshold. This avoids making annotations based on erroneous data	Trimmomatic, sickle, QIIME, mothur, dada2
	Error learning	Certain computer programs use a prediction model to correct reads prior to downstream analyses. Though computationally intensive, this reduces data loss as low-quality parts of reads are corrected instead of simply being removed	dada2
	Sequence merging	Gene sequence fragments are often sequenced from both ends (i.e., paired-end), meaning that each molecule is associated with two reads. Each pair of reads providing from a single DNA fragment must be assembled together prior to continuing the analysis	pandaseq
Data processing	OTU clustering	Merged reads are clustered together, usually on the basis of an identity threshold (often >97%) assuming that similar sequences belong to the same taxonomic entity. Certain methods use exact matching but require the use of an error model (e.g., dada2)	QIIME, mothur, dada2
	Taxonomic assignment	The taxonomic ranks (kingdom, phylum, genus, species) of each OTU are obtained by the consensus of the annotation of all reads used to build it	QIIME, mothur, phyloseq
Statistical analysis	Alpha diversity	Within-sample diversity, i.e., a function of the number of OTUs present in a given condition or sample. For example, the Shannon diversity index measures both richness (the amount of species) and evenness (their distribution)	QIIME, mothur, phyloseq
	Beta diversity	Pairwise distance or dissimilarity between samples. For example, the Bray-Curtis dissimilarity index measures the ratio of unique species versus all species found in a pair of samples. Samples that have a Bray-Curtis index of 1 are entirely composed of mutually exclusive species	QIIME, mothur, phyloseq
	Differential abundance	Involves hypothesis testing to determine which taxa are differentially present in a pair of conditions (or samples)	DEseq2, edgeR (from phyloseq data)
	Co-abundance networks	Built from OTU correlation matrices, those allow the identification of taxa whose abundance is either colinear or mutually exclusive	phyloseq, igraph

Table 4 Major steps in a typical meta-(genomics, transcriptomics) analysis pipeline

Step	Sub-step	Description	Example software
Data preprocessing	Quality filtering and trimming	Removing incorrectly called nucleotides in sequence reads, as well as reads of poor quality. In a trimming approach, reads are trimmed after quality reaches a certain threshold. This avoids making annotations based on erroneous data	Trimmomatic, sickle, Trinity (for metatranscriptomics)
Data processing	De novo assembly	Reconstructing the DNA sequences in the input sample without prior knowledge. Due to the high complexity of microbial community samples, the output data is typically chunks (contigs) corresponding to gene sequences	IDBA-Meta, Ray Meta, SPADES, Trinity (for metatranscriptomics)
	Gene (or transcript) calling	Gene sequences found within contigs are clustered together using an identity threshold, assuming that highly similar sequences are homologous (i.e., code for the same kind of proteins). This process is homologous to the OTU clustering step in metabarcoding	FragGeneScan, Trinity (for metatranscriptomics)
	Annotation	Pairwise alignment of predicted genes against a database of known sequences. One can then predict the nature (function) of proteins encoded by those genes	BLAT, Diamond
Statistical analysis	Metabolic reconstruction	Using the annotation data, one can reconstruct metabolic pathways present in a sample and make comparisons of shared and unique steps between conditions or samples	BlastKOALA
	Differential abundance	Involves hypothesis testing to determine which genes are differentially present (or differentially expressed if metatranscriptomics) in a pair of conditions or samples	DEseq2, edgeR
	Co-abundance networks	Built from abundance correlation matrices, those allow the identification of genes (or transcripts) whose abundance is either colinear or mutually exclusive	phyloseq, igraph

promising to improve the management beneficial microbial functions in aquaculture, as evidenced in other research fields (Culligan et al. 2014). However, the high sequencing depth required to get enough coverage for microbial sequences, the inherent complexity of analyzing metagenomic data (e.g., the lack of functional annotation of most nonhuman microbial transcripts), as well as the required computational power and storage make this approach highly challenging.

Another challenge is that metagenomics gives insight on which gene functions are relevant in a biological system but does not predict their level of activity. Nevertheless, by sequencing host and bacterial messenger RNAs instead of total genomic DNA, one can indeed obtain a *metatranscriptome*, which can be annotated using similar methods; some assembly computer programs even allow the prediction of splicing variants resulting from the transcription of genes with intronic sequences (Haas et al. 2013). Moreover, the transcripts' abundance can be quantified (as it is proportional to the number of mapped reads) making it possible to determine the *level of expression* of a given gene (Bashiardes et al. 2016). Therefore, by allowing to quantify simultaneously both microbial and host tissue gene expression, this methodological approach is thus suitable to shed light on active host microbiota (including pathogens) functional interactions. Accordingly, metatranscriptomics, combined with metagenomic analysis, has shown that, in the human gut, a substantial fraction of microbial transcripts are differentially regulated relatively to their microbial genomic abundances (REF). Though promising, this technique has inherent challenges such as high per-sample cost, depletion of both eukaryotic and bacterial ribosomal RNA transcripts (~90–95% of a total RNA sample), and the lack of standard bioinformatics methods ensuring repeatability across studies (Martin et al. 2018). To our knowledge, no metatranscriptome data of a salmonid has yet been published.

4 Future Perspectives for Microbiota Modulation

4.1 *Host-Microbiota Interactions in Light of the One Health Perspective*

As previously mentioned in chapter "The Rise and Fall of Antibiotics in Aquaculture," an integrated view of biological systems is required to secure aquaculture production and ensure its sustainability. This is exemplified by the One Health perspective, which states that human, animal, and environmental health are co-dependent variables (Lebov et al. 2017). Therefore, acknowledging the contribution of the microbiota in animal health falls within this framework. Roles of the microbiota in salmonid health have been made increasingly clear throughout the last decade (see Sect. 2.2). Two relevant examples that would benefit from a holistic understanding are (1) the emergence of antibiotic resistance in salmonid pathogens, which is worsened by bacteria-to-bacteria gene transfer and the depletion of competing bacteria sensitive to antibiotics (Trudel et al. 2016), and (2) the impact of diet on the microbiota composition and its impact on immune function (see Sect. 2.2).

4.2 *Fine-Grained Modulation Using Dietary Supplements*

4.2.1 Probiotics

Probiotics are defined as a "live microbial culture added to feed or environment to increase viability of the host" (Gram and Ringø 2005). This positive effect on host physiology may originate from several mechanisms of action: (1) mechanisms targeting pathogens such as nutritional competition, diffusion of antimicrobial compounds, or competitive exclusion from epithelial surfaces (Bermudez-Brito et al. 2012; Kamada et al. 2013) and (2) mechanisms targeting the host itself, such as modulation of host immune signaling pathways (Kamada et al. 2013).

Although the specific mechanisms by which some probiotic strains exert their beneficial effects require further investigation, probiotic administration showed promising results on growth performance and general health of salmonid fish (Gatesoupe 2010). Some probiotic candidates showed great promise as prophylactic tools against opportunistic diseases (Boutin et al. 2012; Schubiger et al. 2015; Gauthier et al. 2017a, b). However, few of them, if any, have reached commercialization or even official approval for use in salmonid farming.

We may hope that, in the near future, probiotic administration will be guided by more and more thorough microbiota monitoring studies. For example, if a specific bacterial species is associated with increased immune function, then one or more isolates from this species could be administered to vulnerable fish as a prophylactic treatment. Perhaps microbial community assemblages that are reflective of a good health status will be engineered and administered as "microbiota transplants." The microbiota transplant strategy, for instance, is the most effective therapies (90% efficacy rate) against human nosocomial *Clostridium difficile* infections (Liubakka and Vaughn 2016).

However, the bioengineering of host-associated microbial communities is a complex task, as most of microbial diversity is unculturable using basic microbiological methods (Tanaka et al. 2014). Nevertheless, recent progress in human microbiota studies (Zihler Berner et al. 2013; Auchtung et al. 2015; Dostal et al. 2015) could pave the way toward similar approaches in salmonid health management. For instance, gut bioreactors, a special class of continuous-flow fermenters, proved to be an excellent method to cultivate complex human gut microbiota systems in vitro in highly controlled simulated settings (Macfarlane and Macfarlane 2007).

4.2.2 Prebiotics and Synbiotics

In addition to probiotics, prebiotics are dietary additives that are fermented by the gut microbiota into short-chain fatty acids (SCFAs), which are the main energy source for colonic epithelial cells. The SCFAs also modulate lipid synthesis (Marcil et al. 2002), stimulate the immune system and increase host resistance against pathogens

(Maslowski and Mackay 2010). Synbiotics are combinations of probiotics and prebiotics (Cerezuela et al. 2011). Synbiotics aim to simultaneously seed and maintain probiotic strains as dominant species in the gut. However, despite recent progress, there is limited information available on different aspects of synbiotics effects on fish (Cerezuela et al. 2011; Torrecillas et al. 2018), and their effect on the microbiota composition of salmonids is currently unknown.

4.2.3 Phage Therapy

Because phage particles are very specific to their bacterial hosts, they do not target both pathogens and the normal flora. Furthermore, phage particles replicate at the site of infection; thus curative doses can be fairly small. Moreover, although bacteria can become resistant to phages, these viral organisms can mutate and therefore evolve to counter phage-resistant bacteria (Matsuzaki et al. 2005), which synthetic antimicrobial treatments cannot do. The most important advantage of phages is that they might kill planktonic pathogens living in the surrounding water in addition to pathogens proliferating in carrier fish. Possible drawbacks of phage therapy include the possible transduction of virulence factors between bacteria; in addition, the vertebrate host may mount an immune response against the phage itself (see chapter "Would Bacteriophages Be a New Old Complement to Antibiotics in Aquaculture?" for a detailed discussion on phage therapy in salmonid aquaculture). Candidate phages that infect *A. s. s.* were isolated recently and opened the way to a broad-range treatment against multiple strains of this major salmonid pathogen (Vincent et al. 2017b). Other salmonid pathogens for which phage treatments are under development include *Flavobacterium psychrophilum* (Castillo et al. 2012; Madsen et al. 2013) and *F. columnare* (Prasad et al. 2011).

4.3 *Coarse-Grained Modulation: The Case of K-Selection*

In light of the One Health perspective, one may ponder over the efficacy of microbial management methods. Indeed, the strong demand for fish protein has resulted in a strong pressure on fish farmers to provide for an unprecedented increase of the human population (Duarte et al. 2009). Production strategies that may be efficient from an economical point of view (e.g., maximizing output at a minimal cost) may result in suboptimal rearing conditions such as overcrowding, hypoxia, and handling stress (Heikkinen et al. 2006). Disinfection methods and the aforementioned fine-grained microbiota management methods may help but may not address the root cause of certain diseases (i.e., poor rearing conditions). Current microbial management strategies aim to reduce the microbial load (e.g., surface disinfection of eggs and UV irradiation of incoming water) by assuming that fewer microbes translate into fewer risks of infectious disease. By doing so, those methods actually increase the water's carrying capacity for generalist opportunistic microbes (r-strategists) whose life strategy is to proliferate as rapidly as possible, sometimes at the expense of their host (i.e., opportunistic pathogens) (Vadstein et al. 2018).

As a matter of fact, cod larvae (*Gadus morhua*) reared in recirculating aquaculture systems (RAS) had 72% higher survival rates than larvae reared in flow-through aquaculture (FTS) systems (Attramadal et al. 2014). RAS systems are specifically designed to promote the establishment of a complex microbial ecosystem that feeds of dissolved organic compounds and waste produced by the fish. Moreover, those systems are self-sustained (water is recirculating), resulting in a mature, specialist microbial community (K-strategists) and, therefore, low carrying capacity for opportunistic r-strategists. K-selection of microbial communities holds great promise as a microbiota management tool to promote fish health [for more details on K-selection, see chapter "Controlling Factors for Community Assembly in Developing Cod Larvae (*Gadus morhua*)"]. To our knowledge, research on K-selection of microbial communities in aquaculture has not yet been performed on salmonid fish.

4.4 Toward Real-Time Microbiota Monitoring

Novel DNA sequencing technologies have raised the bar in terms of throughput and scalability. For example, Oxford Nanopore has launched the MinION system, which allows amplicon and metagenome sequencing on a USB stick-side apparatus plugged on a laptop computer (Krehenwinkel et al. 2018). However, its high incorrect basecalling rate of 3%, which is equal to the well-adopted difference threshold for clustering OTUs, limits its usage in metabarcoding studies (Kerkhof et al. 2017). Because single-nucleotide differences are often critical to resolve distinct genera, it is imperative to distinguish biological sequence variation from amplicon sequencing errors (Callahan et al. 2016). With further improvements in sequencing accuracy and novel analysis pipelines, near-instant visualization and analysis of complex microbiome data from a laptop computer will perhaps be possible in the near future.

5 Conclusion

Since 2010, salmonids are second to carps as the most important group of farmed fish, with a total output of over 2 million tonnes. This results in a strong pressure on fish farmers to keep up with high demand for this source of animal protein. Intensive farming practices are commonly used in order to compensate for the immune impairment that results from an over-elicited stress response. This can disrupt global interactions between the host and its microbial symbionts (i.e., microbiota) that play a key role in maintaining fish health in the long term. This chapter presented an overview of the current knowledge on the taxonomic composition (i.e., diversity and structure) of salmonid microbiota and the state of the art on microbial profiling and modulation, as well as current research gaps and perspectives. A special attention has been given to the microbiota of salmonids that are significantly important in aquaculture.

- Atlantic salmon has the most extensively characterized microbiota of all salmonids to this present day, with 17 dedicated studies published between 2007 and 2018. Most of this research focused on the skin and gut microbiota, including assessments of its response to migration, nutrition, antibiotherapy, and captivity.
- Rainbow trout is the second most produced salmonid fish in salmonid aquaculture and is also the second most studied in terms of host-microbiota interactions. A total of 14 studies investigated the influence of diet, growth promotion, and pathogen inhibition, as well as diet-immunity interactions and their impact on the rainbow trout microbiota. However, none of those studies addressed the impact of antibiotherapy on the microbiota composition and subsequent effects on fish health.
- The brook charr microbiota remains largely mischaracterized, except for the skin mucus microbiota, for which response to intensive rearing conditions and symbiont-pathogen interactions (including interindividual variations) were investigated. To our knowledge, no published studies have yet discussed the impact of diet or therapeutic tools on both microbiota structure and brook charr physiology, though ongoing studies are currently underway.

High-resolution study of the microbiota has been made possible with the advent of high-throughput DNA sequencing during the late 2000s. Several methods have emerged, each offering distinct elements of information on the host-microbiota complex.

- Whole-genome shotgun (WGS) sequencing aims to obtain the whole DNA sequence of a specific organism.
- Metabarcoding is the massively parallel sequencing of a universal genetic marker to infer the taxonomic census of a community. It is currently the most common method in modern microbiome studies, regardless of which organisms are studied.
- Metagenomics is the whole-genome shotgun sequencing of total DNA extracts from complex communities instead of a single organism's DNA, leading to the untargeted sequencing of all microorganisms' genomic sequences present in a given sample, in addition to the host organism's genome.

These methods allowed the study of salmonid microbiota from various levels of organization, i.e., from an individual microbe's genome to the global functional interactions between host and the hundreds of microbial symbionts. Several health-promoting and therapeutic applications have benefitted from those omics-based methods. Whole-genome sequencing allows a thorough characterization of some bacterial probiotic candidates, as well as bacteriophages, that show great promise as prophylactic tools against opportunistic diseases. Metabarcoding and metagenomics allow modelling the microbiota structure as a function of health and treatment parameters. The latter are particularly useful in monitoring difference in microbiota composition across conditions that tremendously differ between each other (e.g., flow-through aquaculture vs recirculating aquaculture systems).

However, there are significant knowledge gaps in how salmonid-microbiota systems are affected as a whole by antibiotherapy, diet, and rearing conditions. The cause-effect relationships between treatments (or conditions), differential microbiota composition, and host physiology often remain unresolved in most studies. In addition, there are great inherent difficulties associated with the analysis of such a complex multivariate system as the microbiota, coupled with the lack of methodological consensus between studies. Perhaps the implementation of "good practices" and new technological advances should help resolving the complexity of host-microbiota systems to the benefit of salmonid aquaculture. Finally, there is growing awareness that fish and their rearing environments are complex ecosystems and, accordingly, that the well-being of both fish and microbiota must be considered when developing therapeutic tools or intensive rearing protocols.

References

Araújo C, Muñoz-Atienza E, Nahuelquín Y et al (2015) Inhibition of fish pathogens by the microbiota from rainbow trout (*Oncorhynchus mykiss*, Walbaum) and rearing environment. Anaerobe 32:7–14. https://doi.org/10.1016/j.anaerobe.2014.11.001

Aßhauer KP, Wemheuer B, Daniel R, Meinicke P (2015) Tax4Fun: predicting functional profiles from metagenomic 16S rRNA data. Bioinformatics 31:2882–2884. https://doi.org/10.1093/bioinformatics/btv287

Attramadal KJK, Truong TMH, Bakke I et al (2014) RAS and microbial maturation as tools for K-selection of microbial communities improve survival in cod larvae. Aquaculture 432:483–490. https://doi.org/10.1016/j.aquaculture.2014.05.052

Auchtung JM, Robinson CD, Britton RA (2015) Cultivation of stable, reproducible microbial communities from different fecal donors using minibioreactor arrays (MBRAs). Microbiome 3. https://doi.org/10.1186/s40168-015-0106-5

Balassiano IT, Bastos Mdo C, Madureira DJ et al (2007) The involvement of *tetA* and *tetE* tetracycline resistance genes in plasmid and chromosomal resistance of *Aeromonas* in Brazilian strains. Mem Inst Oswaldo Cruz 102:861–866. https://doi.org/10.1590/S0074-02762007005000121

Barton BA (2002) Stress in fishes: a diversity of responses with particular reference to changes in circulating corticosteroids. Integr Comp Biol 42:517–525. https://doi.org/10.1093/icb/42.3.517

Bashiardes S, Zilberman-Schapira G, Elinav E (2016) Use of metatranscriptomics in microbiome research. Bioinform Biol Insights 10:19–25. https://doi.org/10.4137/BBI.S34610

Bermudez-Brito M, Plaza-Díaz J, Muñoz-Quezada S et al (2012) Probiotic mechanisms of action. Ann Nutr Metab 61:160–174. https://doi.org/10.1159/000342079

Boutin S, Bernatchez L, Audet C, Derôme N (2012) Antagonistic effect of indigenous skin bacteria of brook charr (*Salvelinus fontinalis*) against *Flavobacterium columnare* and *F. psychrophilum*. Vet Microbiol 155:355–361. https://doi.org/10.1016/j.vetmic.2011.09.002

Boutin S, Audet C, Derôme N (2013a) Probiotic treatment by indigenous bacteria decreases mortality without disturbing the natural microbiota of *Salvelinus fontinalis*. Can J Microbiol 59:662–670

Boutin S, Bernatchez L, Audet C, Derôme N (2013b) Network analysis highlights complex interactions between pathogen, host and commensal microbiota. PLoS One 8:e84772. https://doi.org/10.1371/journal.pone.0084772

Boutin S, Sauvage C, Bernatchez L et al (2014) Inter individual variations of the fish skin microbiota: host genetics basis of mutualism? PLoS One 9:e102649. https://doi.org/10.1371/journal.pone.0102649

Bullock G, Cipriano R, Snieszko S (1983) Furunculosis and other diseases caused by *Aeromonas salmonicida*. US Fish & Wildlife Publications, Washington, DC

Callahan BJ, McMurdie PJ, Rosen MJ et al (2016) DADA2: high resolution sample inference from Illumina amplicon data. Nat Methods 13:581–583. https://doi.org/10.1038/nmeth.3869

Castillo D, Higuera G, Villa M et al (2012) Diversity of *Flavobacterium psychrophilum* and the potential use of its phages for protection against bacterial cold water disease in salmonids. J Fish Dis 35:193–201. https://doi.org/10.1111/j.1365-2761.2011.01336.x

Ceppa F, Faccenda F, Filippo CD et al (2018) Influence of essential oils in diet and life-stage on gut microbiota and fillet quality of rainbow trout (*Oncorhynchus mykiss*). Int J Food Sci Nutr 69:318–333. https://doi.org/10.1080/09637486.2017.1370699

Cerezuela R, Meseguer J, Esteban M (2011) Current knowledge in synbiotic use for fish aquaculture: a review. J Aquac Res Dev 1:1–7

Chen L, Zheng D, Liu B et al (2016) VFDB 2016: hierarchical and refined dataset for big data analysis—10 years on. Nucleic Acids Res 44:D694–D697. https://doi.org/10.1093/nar/gkv1239

Cherrington CA, Hinton M, Pearson GR, Chopra I (1991) Short-chain organic acids at pH 5.0 kill *Escherichia coli* and *Salmonella* spp. without causing membrane perturbation. J Appl Bacteriol 70:161–165

Cole JR, Wang Q, Fish JA et al (2014) Ribosomal Database Project: data and tools for high throughput rRNA analysis. Nucleic Acids Res 42:D633–D642. https://doi.org/10.1093/nar/gkt1244

Culligan EP, Sleator RD, Marchesi JR, Hill C (2014) Metagenomics and novel gene discovery: promise and potential for novel therapeutics. Virulence 5:399–412. https://doi.org/10.4161/viru.27208

Cummings JH (1984) Cellulose and the human gut. Gut 25:805–810. https://doi.org/10.1136/gut.25.8.805

Dehler CE, Secombes CJ, Martin SAM (2017) Seawater transfer alters the intestinal microbiota profiles of Atlantic salmon (*Salmo salar* L.). Sci Rep 7:13877. https://doi.org/10.1038/s41598-017-13249-8

Delsuc F, Brinkmann H, Philippe H (2005) Phylogenomics and the reconstruction of the tree of life. Nat Rev Genet 6:361–375. https://doi.org/10.1038/nrg1603

Derome N, Duchesne P, Bernatchez L (2006) Parallelism in gene transcription among sympatric lake whitefish (*Coregonus clupeaformis Mitchill*) ecotypes. Mol Ecol 15:1239–1249. https://doi.org/10.1111/j.1365-294X.2005.02968.x

Desai AR, Links MG, Collins SA et al (2012) Effects of plant-based diets on the distal gut microbiome of rainbow trout (*Oncorhynchus mykiss*). Aquaculture 350–353:134–142. https://doi.org/10.1016/j.aquaculture.2012.04.005

DeSantis TZ, Hugenholtz P, Larsen N et al (2006) Greengenes, a chimera-checked 16S rRNA gene database and workbench compatible with ARB. Appl Environ Microbiol 72:5069–5072. https://doi.org/10.1128/AEM.03006-05

Dostal A, Lacroix C, Bircher L et al (2015) Iron modulates butyrate production by a child gut microbiota in vitro. MBio 6:e01453–e01415. https://doi.org/10.1128/mBio.01453-15

Duarte CM, Holmer M, Olsen Y et al (2009) Will the oceans help feed humanity? https://doi.org/10.1525/bio.2009.59.11.8. http://www.bioone.org/doi/abs/10.1525/bio.2009.59.11.8. Accessed 18 Dec 2018

FAO Fisheries & Aquaculture (2018) FAO fisheries & aquaculture – cultured aquatic species information programme – *Oncorhynchus mykiss* (Walbaum, 1792). FAO Fisheries &

Aquaculture. http://www.fao.org/fishery/culturedspecies/Oncorhynchus_mykiss/en. Accessed 18 Dec 2018

FAO/OMS (2015) An overview of Atlantic salmon, its natural history, aquaculture, and genetic engineering. http://www.fda.gov/AdvisoryCommittees/CommitteesMeetingMaterials/VeterinaryMedicineAdvisoryCommittee/ucm222635.htm#Section_3:_Aquaculture. Accessed 15 Jun 2015

Fast MD, Hosoya S, Johnson SC, Afonso LOB (2008) Cortisol response and immune-related effects of Atlantic salmon (*Salmo salar* Linnaeus) subjected to short- and long-term stress. Fish Shellfish Immunol 24:194–204. https://doi.org/10.1016/j.fsi.2007.10.009

Freschi L, Vincent AT, Jeukens J et al (2018) The *Pseudomonas aeruginosa* pan-genome provides new insights on its population structure, horizontal gene transfer and pathogenicity. Genome Biol Evol. https://doi.org/10.1093/gbe/evy259

Gajardo K, Rodiles A, Kortner TM et al (2016) A high-resolution map of the gut microbiota in Atlantic salmon (*Salmo salar*): a basis for comparative gut microbial research. Sci Rep 6:30893. https://doi.org/10.1038/srep30893

Gajardo K, Jaramillo-Torres A, Kortner TM et al (2017) Alternative protein sources in the diet modulate microbiota and functionality in the distal intestine of Atlantic Salmon (*Salmo salar*). Appl Environ Microbiol 83. https://doi.org/10.1128/AEM.02615-16

Garrido-Cardenas JA, Manzano-Agugliaro F (2017) The metagenomics worldwide research. Curr Genet 63:819–829. https://doi.org/10.1007/s00294-017-0693-8

Gatesoupe FJ (2010) Chapter 32 – Probiotics and other microbial manipulations in fish feeds: prospective health benefits. In: Watson RR, Preedy VR (eds) Bioactive foods in promoting health. Academic Press, Boston, pp 541–552

Gauthier J (2016) Développement d'outils thérapeutiques à base de probiotiques endogènes contre la furonculose (*Aeromonas salmonicida*) chez l'Omble de fontaine d'aquaculture (Phase 1 – in vitro)

Gauthier J, Charette SJ, Derome N (2017a) Draft genome sequence of *Pseudomonas fluorescens* ML11A, an endogenous strain from Brook Charr with antagonistic properties against *Aeromonas salmonicida subsp. salmonicida*. Genome Announc 5. https://doi.org/10.1128/genomeA.01716-16

Gauthier J, Vincent AT, Charette SJ, Derome N (2017b) Strong genomic and phenotypic heterogeneity in the *Aeromonas sobria* species complex. Front Microbiol 8. https://doi.org/10.3389/fmicb.2017.02434

Gauthier J, Vincent AT, Charette SJ, Derome N (2018) A brief history of bioinformatics. Brief Bioinform. https://doi.org/10.1093/bib/bby063

Geurden I, Mennigen J, Plagnes-Juan E et al (2014) High or low dietary carbohydrate:protein ratios during first-feeding affect glucose metabolism and intestinal microbiota in juvenile rainbow trout. J Exp Biol 217:3396–3406. https://doi.org/10.1242/jeb.106062

Gram L, Ringø E (2005) Prospects of fish probiotics. In: Holzapfel WH, Naughton PJ (eds) Microbial ecology of the growing animal. Elsevier, Amsterdam, pp 379–417

Haas BJ, Papanicolaou A, Yassour M et al (2013) De novo transcript sequence reconstruction from RNA-Seq: reference generation and analysis with Trinity. Nat Protoc 8. https://doi.org/10.1038/nprot.2013.084

Hammami R, Fernandez B, Lacroix C, Fliss I (2013) Anti-infective properties of bacteriocins: an update. Cell Mol Life Sci 70:2947–2967. https://doi.org/10.1007/s00018-012-1202-3

Harvey AC, Solberg MF, Troianou E et al (2016) Plasticity in growth of farmed and wild Atlantic salmon: is the increased growth rate of farmed salmon caused by evolutionary adaptations to the commercial diet? BMC Evol Biol 16:264. https://doi.org/10.1186/s12862-016-0841-7

He X, Chaganti SR, Heath DD (2018) Population-specific responses to interspecific competition in the gut microbiota of two Atlantic Salmon (*Salmo salar*) populations. Microb Ecol 75:140–151. https://doi.org/10.1007/s00248-017-1035-6

Heikkinen J, Vielma J, Kemiläinen O et al (2006) Effects of soybean meal based diet on growth performance, gut histopathology and intestinal microbiota of juvenile rainbow trout

(*Oncorhynchus mykiss*). Aquaculture 261:259–268. https://doi.org/10.1016/j.aquaculture.2006.07.012

Herman EM, Schmidt MA (2016) The potential for engineering enhanced functional-feed soybeans for sustainable aquaculture feed. Front Plant Sci 7. https://doi.org/10.3389/fpls.2016.00440

Héry M, Volant A, Garing C et al (2014) Diversity and geochemical structuring of bacterial communities along a salinity gradient in a carbonate aquifer subject to seawater intrusion. FEMS Microbiol Ecol 90:922–934. https://doi.org/10.1111/1574-6941.12445

Ingerslev H-C, Strube ML, Jørgensen L v G et al (2014) Diet type dictates the gut microbiota and the immune response against *Yersinia ruckeri* in rainbow trout (*Oncorhynchus mykiss*). Fish Shellfish Immunol 40:624–633. https://doi.org/10.1016/j.fsi.2014.08.021

Jaafar RM, Kania PW, Larsen AH et al (2013) Gut microbiota changes in rainbow trout, *Oncorhynchus mykiss* (Walbaum), during organic acid feed supplementation and *Yersinia ruckeri* infection. J Fish Dis 36:599–606. https://doi.org/10.1111/jfd.12047

Jia B, Raphenya AR, Alcock B et al (2017) CARD 2017: expansion and model-centric curation of the comprehensive antibiotic resistance database. Nucleic Acids Res 45:D566–D573. https://doi.org/10.1093/nar/gkw1004

Johnson SC, Albright LJ (1992) Effects of cortisol implants on the susceptibility and the histopathology of the responses of naive coho salmon *Oncorhynchus kisutch* to experimental infection with *Lepeophtheirus salmonis* (Copepoda: Caligidae). Dis Aquat Org 14:195–205

Ju F, Zhang T (2015) 16S rRNA gene high-throughput sequencing data mining of microbial diversity and interactions. Appl Microbiol Biotechnol 99:4119–4129. https://doi.org/10.1007/s00253-015-6536-y

Kadlec K, von Czapiewski E, Kaspar H et al (2011) Molecular basis of sulfonamide and trimethoprim resistance in fish-pathogenic *Aeromonas* isolates. Appl Environ Microbiol 77:7147–7150. https://doi.org/10.1128/AEM.00560-11

Kamada N, Chen GY, Inohara N, Núñez G (2013) Control of pathogens and pathobionts by the gut microbiota. Nat Immunol 14:685–690. https://doi.org/10.1038/ni.2608

Kanehisa M, Sato Y, Morishima K (2016) BlastKOALA and GhostKOALA: KEGG tools for functional characterization of genome and metagenome sequences. J Mol Biol 428:726–731. https://doi.org/10.1016/j.jmb.2015.11.006

Karlsson F, Tremaroli V, Nielsen J, Bäckhed F (2013) Assessing the human gut microbiota in metabolic diseases. Diabetes 62:3341–3349. https://doi.org/10.2337/db13-0844

Kerkhof LJ, Dillon KP, Häggblom MM, McGuinness LR (2017) Profiling bacterial communities by MinION sequencing of ribosomal operons. Microbiome 5:116. https://doi.org/10.1186/s40168-017-0336-9

Konstantinidis KT, Tiedje JM (2005) Genomic insights that advance the species definition for prokaryotes. Proc Natl Acad Sci U S A 102:2567–2572. https://doi.org/10.1073/pnas.0409727102

Krehenwinkel H, Pomerantz A, Henderson JB et al (2018) Nanopore sequencing of long ribosomal DNA amplicons enables portable and simple biodiversity assessments with high phylogenetic resolution across broad taxonomic scale. bioRxiv 358572. https://doi.org/10.1101/358572

Langille MGI, Zaneveld J, Caporaso JG et al (2013) Predictive functional profiling of microbial communities using 16S rRNA marker gene sequences. Nat Biotechnol 31:814–821. https://doi.org/10.1038/nbt.2676

Lavoie C, Courcelle M, Redivo B, Derome N (2018) Structural and compositional mismatch between captive and wild Atlantic salmon (*Salmo salar*) parrs' gut microbiota highlights the relevance of integrating molecular ecology for management and conservation methods. Evol Appl 11:1671–1685. https://doi.org/10.1111/eva.12658

Lebov J, Grieger K, Womack D et al (2017) A framework for one health research. One Health 3:44–50. https://doi.org/10.1016/j.onehlt.2017.03.004

Liubakka A, Vaughn BP (2016) *Clostridium difficile* infection and fecal microbiota transplant. AACN Adv Crit Care 27:324–337. https://doi.org/10.4037/aacnacc2016703

Llewellyn MS, Boutin S, Hoseinifar SH, Derome N (2014) Teleost microbiomes: the state of the art in their characterization, manipulation and importance in aquaculture and fisheries. Front Microbiol 5:207. https://doi.org/10.3389/fmicb.2014.00207

Lokesh J, Kiron V (2016) Transition from freshwater to seawater reshapes the skin-associated microbiota of Atlantic salmon. Sci Rep 6:19707. https://doi.org/10.1038/srep19707

Macfarlane GT, Macfarlane S (2007) Models for intestinal fermentation: association between food components, delivery systems, bioavailability and functional interactions in the gut. Curr Opin Biotechnol 18:156–162. https://doi.org/10.1016/j.copbio.2007.01.011

Madaro A, Olsen RE, Kristiansen TS et al (2015) Stress in Atlantic salmon: response to unpredictable chronic stress. J Exp Biol 218:2538–2550. https://doi.org/10.1242/jeb.120535

Madsen L, Bertelsen SK, Dalsgaard I, Middelboe M (2013) Dispersal and survival of *Flavobacterium psychrophilum* phages in vivo in rainbow trout and in vitro under laboratory conditions: implications for their use in phage therapy. Appl Environ Microbiol 79:4853–4861. https://doi.org/10.1128/AEM.00509-13

Mahato NK, Gupta V, Singh P et al (2017) Microbial taxonomy in the era of OMICS: application of DNA sequences, computational tools and techniques. Antonie Van Leeuwenhoek 110:1357–1371. https://doi.org/10.1007/s10482-017-0928-1

Marcil V, Delvin E, Seidman E et al (2002) Modulation of lipid synthesis, apolipoprotein biogenesis, and lipoprotein assembly by butyrate. Am J Physiol Gastrointest Liver Physiol 283:G340–G346

Martin TC, Visconti A, Spector TD, Falchi M (2018) Conducting metagenomic studies in microbiology and clinical research. Appl Microbiol Biotechnol 102:8629–8646. https://doi.org/10.1007/s00253-018-9209-9

Maslowski KM, Mackay CR (2010) Diet, gut microbiota and immune responses. Nat Immunol 12:5–9

Matsuzaki S, Rashel M, Uchiyama J et al (2005) Bacteriophage therapy: a revitalized therapy against bacterial infectious diseases. J Infect Chemother 11:211–219

Maule AG, Tripp RA, Kaattari SL, Schreck CB (1989) Stress alters immune function and disease resistance in Chinook salmon (*Oncorhynchus tshawytscha*). J Endocrinol 120:135–142

Mayer EA (2000) The neurobiology of stress and gastrointestinal disease. Gut 47:861–869. https://doi.org/10.1136/gut.47.6.861

Méndez J, Guijarro JA (2013) In vivo monitoring of *Yersinia ruckeri* in fish tissues: progression and virulence gene expression. Environ Microbiol Rep 5:179–185. https://doi.org/10.1111/1758-2229.12030

Milot E, Perrier C, Papillon L et al (2013) Reduced fitness of Atlantic salmon released in the wild after one generation of captive breeding. Evol Appl 6:472–485. https://doi.org/10.1111/eva.12028

Miranda CD, Zemelman R (2002) Bacterial resistance to oxytetracycline in Chilean salmon farming. Aquaculture 212:31–47. https://doi.org/10.1016/S0044-8486(02)00124-2

Morey JN, Boggero IA, Scott AB, Segerstrom SC (2015) Current directions in stress and human immune function. Curr Opin Psychol 5:13–17. https://doi.org/10.1016/j.copsyc.2015.03.007

Navarrete P, Mardones P, Opazo R et al (2008) Oxytetracycline treatment reduces bacterial diversity of intestinal microbiota of Atlantic salmon. J Aquat Anim Health 20:177–183. https://doi.org/10.1577/H07-043.1

Naylor RL, Goldburg RJ, Primavera JH et al (2000) Effect of aquaculture on world fish supplies. Nature 405:1017–1024. https://doi.org/10.1038/35016500

Newaj-Fyzul A, Austin B (2015) Probiotics, immunostimulants, plant products and oral vaccines, and their role as feed supplements in the control of bacterial fish diseases. J Fish Dis 38:937–955. https://doi.org/10.1111/jfd.12313

Nordmo, Riseth H, Varma et al (1998) Evaluation of florfenicol in Atlantic salmon, *Salmo salar* L.: efficacy against furunculosis due to *Aeromonas salmonicida* and cold water vibriosis due to *Vibrio salmonicida*. J Fish Dis 21:289–297. https://doi.org/10.1046/j.1365-2761.1998.00106.x

Osman M-A, Neoh H-M, Ab Mutalib N-S et al (2018) 16S rRNA gene sequencing for deciphering the colorectal cancer gut microbiome: current protocols and workflows. Front Microbiol 9:767. https://doi.org/10.3389/fmicb.2018.00767

Overbeek R, Olson R, Pusch GD et al (2014) The SEED and the rapid annotation of microbial genomes using subsystems technology (RAST). Nucleic Acids Res 42:D206–D214. https://doi.org/10.1093/nar/gkt1226

Panek M, Paljetak HČ, Barešić A et al (2018) Methodology challenges in studying human gut microbiota – effects of collection, storage, DNA extraction and next generation sequencing technologies. Sci Rep 8:5143. https://doi.org/10.1038/s41598-018-23296-4

Park H, Weier S, Razvi F et al (2017) Towards the development of a sustainable soya bean-based feedstock for aquaculture. Plant Biotechnol J 15:227–236. https://doi.org/10.1111/pbi.12608

Pickering AD, Duston J (1982) Administration of cortisol to brown trout, *Salmo trutta* L., and its effects on the susceptibility to *Saprolegnia* infection and furunculosis. J Fish Biol 23:163–175

Pickering AD, Pottinger TG (1985) Cortisol can increase the susceptibility of brown trout, *Salmo trutta* L., to disease without reducing the white blood cell count. J Fish Biol 27:611–619. https://doi.org/10.1111/j.1095-8649.1985.tb03206.x

Pollock J, Glendinning L, Wisedchanwet T, Watson M (2018) The madness of microbiome: attempting to find consensus "best practice" for 16s microbiome studies. Appl Environ Microbiol 84:e02627–e02617. https://doi.org/10.1128/AEM.02627-17

Prasad Y, Arpana, Kumar D, Sharma AK (2011) Lytic bacteriophages specific to *Flavobacterium columnare* rescue catfish, *Clarias batrachus* (Linn.) from columnaris disease. J Environ Biol 32:161–168

Qin J, Li R, Raes J et al (2010) A human gut microbial gene catalogue established by metagenomic sequencing. Nature 464:59–65. https://doi.org/10.1038/nature08821

Quince C, Walker AW, Simpson JT et al (2017) Shotgun metagenomics, from sampling to analysis. Nat Biotechnol 35:833–844. https://doi.org/10.1038/nbt.3935

Raida MK, Larsen JL, Nielsen ME, Buchmann K (2003) Enhanced resistance of rainbow trout, *Oncorhynchus mykiss* (Walbaum), against *Yersinia ruckeri* challenge following oral administration of *Bacillus subtilis* and *B. licheniformis* (BioPlus2B). J Fish Dis 26:495–498. https://doi.org/10.1046/j.1365-2761.2003.00480.x

Reith ME, Singh RK, Curtis B et al (2008) The genome of *Aeromonas salmonicida subsp. salmonicida* A449: insights into the evolution of a fish pathogen. BMC Genomics 9:427. https://doi.org/10.1186/1471-2164-9-427

Rimoldi S, Terova G, Ascione C et al (2018) Next generation sequencing for gut microbiome characterization in rainbow trout (*Oncorhynchus mykiss*) fed animal by-product meals as an alternative to fishmeal protein sources. PLoS One 13:e0193652. https://doi.org/10.1371/journal.pone.0193652

Rochat T, Barbier P, Nicolas P et al (2017) Complete genome sequence of *Flavobacterium psychrophilum* strain OSU THCO2-90, used for functional genetic analysis. Genome Announc 5. https://doi.org/10.1128/genomeA.01665-16

Rudi K, Angell IL, Pope PB et al (2018) Stable core gut microbiota across the freshwater-to-saltwater transition for farmed Atlantic Salmon. Appl Environ Microbiol 84:e01974–e01917. https://doi.org/10.1128/AEM.01974-17

Schreck CB (1982) Stress and rearing of salmonids. Aquaculture 28:241–249. https://doi.org/10.1016/0044-8486(82)90026-6

Schubiger CB, Orfe LH, Sudheesh PS et al (2015) Entericidin is required for a probiotic treatment (*Enterobacter* sp. strain C6-6) to protect trout from cold-water disease challenge. Appl Environ Microbiol 81:658–665. https://doi.org/10.1128/AEM.02965-14

Sender R, Fuchs S, Milo R (2016) Revised estimates for the number of human and bacteria cells in the body. bioRxiv 036103. https://doi.org/10.1101/036103

Silby MW, Winstanley C, Godfrey SAC et al (2011) *Pseudomonas* genomes: diverse and adaptable. FEMS Microbiol Rev 35:652–680. https://doi.org/10.1111/j.1574-6976.2011.00269.x

Swiatczak B, Cohen IR (2015) Gut feelings of safety: tolerance to the microbiota mediated by innate immune receptors. Microbiol Immunol 59:573–585. https://doi.org/10.1111/1348-0421.12318

Tagini F, Greub G (2017) Bacterial genome sequencing in clinical microbiology: a pathogen-oriented review. Eur J Clin Microbiol Infect Dis 36:2007–2020. https://doi.org/10.1007/s10096-017-3024-6

Tanaka T, Kawasaki K, Daimon S et al (2014) A hidden pitfall in the preparation of agar media undermines microorganism cultivability. Appl Environ Microbiol 80:7659–7666. https://doi.org/10.1128/AEM.02741-14

Tatusova T, DiCuccio M, Badretdin A et al (2016) NCBI prokaryotic genome annotation pipeline. Nucleic Acids Res 44:6614–6624. https://doi.org/10.1093/nar/gkw569

Tobback E, Decostere A, Hermans K et al (2007) *Yersinia ruckeri* infections in salmonid fish. J Fish Dis 30:257–268. https://doi.org/10.1111/j.1365-2761.2007.00816.x

Torrecillas S, Rivero-Ramírez F, Izquierdo MS et al (2018) Feeding European sea bass (*Dicentrarchus labrax*) juveniles with a functional synbiotic additive (mannan oligosaccharides and *Pediococcus acidilactici*): an effective tool to reduce low fishmeal and fish oil gut health effects? Fish Shellfish Immunol 81:10–20. https://doi.org/10.1016/j.fsi.2018.07.007

Trudel MV, Vincent AT, Attéré SA et al (2016) Diversity of antibiotic-resistance genes in Canadian isolates of *Aeromonas salmonicida* subsp. *salmonicida*: dominance of pSN254b and discovery of pAsa8. Sci Rep 6. https://doi.org/10.1038/srep35617

Vadstein O, Attramadal KJK, Bakke I, Olsen Y (2018) K-selection as microbial community management strategy: a method for improved viability of larvae in aquaculture. Front Microbiol 9. https://doi.org/10.3389/fmicb.2018.02730

Vázquez-Baeza Y, Hyde ER, Suchodolski JS, Knight R (2016) Dog and human inflammatory bowel disease rely on overlapping yet distinct dysbiosis networks. Nat Microbiol 1:16177. https://doi.org/10.1038/nmicrobiol.2016.177

Vincent AT, Derome N, Boyle B et al (2017a) Next-generation sequencing (NGS) in the microbiological world: how to make the most of your money. J Microbiol Methods 138:60–71. https://doi.org/10.1016/j.mimet.2016.02.016

Vincent AT, Paquet VE, Bernatchez A et al (2017b) Characterization and diversity of phages infecting *Aeromonas salmonicida* subsp. *salmonicida*. Sci Rep 7:7054. https://doi.org/10.1038/s41598-017-07401-7

Watson RA, Pollack JB (2001) Symbiotic composition and evolvability. In: Kelemen J, Sosík P (eds) Advances in artificial life. Springer, Berlin, pp 480–490

Wilk R, Andersen K, Uglenes I, Egidius E (1989) Cortisol-induced increase in susceptibility of Atlantic salmon, *Salmo salar*, to *Vibrio salmonicida*, together with effects on the blood cell pattern. Aquaculture 83:201–215

Wong S, Waldrop T, Summerfelt S et al (2013) Aquacultured rainbow trout (*Oncorhynchus mykiss*) possess a large core intestinal microbiota that is resistant to variation in diet and rearing density. Appl Environ Microbiol 79:4974–4984. https://doi.org/10.1128/AEM.00924-13

Yilmaz P, Parfrey LW, Yarza P et al (2014) The SILVA and "All-species Living Tree Project (LTP)" taxonomic frameworks. Nucleic Acids Res 42:D643–D648. https://doi.org/10.1093/nar/gkt1209

Zepeda Mendoza ML, Sicheritz-Pontén T, Gilbert MTP (2015) Environmental genes and genomes: understanding the differences and challenges in the approaches and software for their analyses. Brief Bioinform 16:745–758. https://doi.org/10.1093/bib/bbv001

Zihler Berner A, Fuentes S, Dostal A et al (2013) Novel Polyfermentor intestinal model (PolyFermS) for controlled ecological studies: validation and effect of pH. PLoS One 8:e77772. https://doi.org/10.1371/journal.pone.0077772

Would Bacteriophages Be a New Old Complement to Antibiotics in Aquaculture?

Antony T. Vincent, Valérie E. Paquet, Sylvain Moineau, and Steve J. Charette

Abstract The rise of antibiotic-resistant bacterial strains is a global concern in many sectors, such as aquaculture, as described in chapter "The Rise and Fall of Antibiotics in Aquaculture." To counter this phenomenon, several alternatives or complement to antibiotics have been investigated. Here, we will look at one of those proposed strategies that of using bacteria-specific viruses, called bacteriophages, or commonly phages. Since their discovery in the early 1900s, bacteriophage treatments have had a fleeting popularity in Western countries due to several scientific reasons as well as in some cases, political motives. Only recently, with the appearance of multidrug-resistant bacterial strains, a new craze for phage therapy appeared in Western countries. In an aquaculture context, some studies have shown promising results for the treatment of fish diseases using phages. More specifically, the experimentations with phage cocktail against *A. salmonicida*, infectious agent of furunculosis in salmonids, both in vitro and in vivo, provide an interesting foundation for future alternative treatments. However, since phages and bacteria are

A. T. Vincent
INRS-Institut Armand-Frappier, Bacterial Symbionts Evolution, Laval City, QC, Canada

V. E. Paquet · S. J. Charette (✉)
Institut de Biologie Intégrative et des Systèmes (IBIS), Université Laval, Quebec City, QC, Canada

Département de biochimie, de microbiologie et de bio-informatique, Faculté des sciences et de génie, Université Laval, Quebec City, QC, Canada

Centre de recherche de l'Institut universitaire de cardiologie et de pneumologie de Québec (IUCPQ), Quebec City, QC, Canada
e-mail: Steve.charette@bcm.ulaval.ca

S. Moineau
Département de biochimie, de microbiologie et de bio-informatique, Faculté des sciences et de génie, Université Laval, Quebec City, QC, Canada

Groupe de recherche en écologie buccale (GREB), Faculté de médecine dentaire, Université Laval, Quebec City, QC, Canada

Félix d'Hérelle Reference Center for Bacterial Viruses, Faculté de médecine dentaire, Université Laval, Quebec City, QC, Canada

N. Derome (ed.), *Microbial Communities in Aquaculture Ecosystems*,
https://doi.org/10.1007/978-3-030-16190-3_3

evolving entities, this biological war is far from over. The presence of phage-resistance mechanisms in bacteria and other technical aspects of phage therapy in aquaculture are factors to consider before having any applicable treatments.

1 The Enemy of My Enemy Is My Friend

Viruses are biological entities capable of infecting the entire kingdom of life, including bacteria. Bacterial viruses are called bacteriophages or more commonly phages, and they were co-discovered independently by Frederick Twort (1915) and Félix d'Hérelle (1917). Rapidly after their discovery, studies on the therapeutic potential of phages, also called phagotherapy, were published (McKinley 1923). Félix d'Hérelle was the pioneer of the phagotherapy, using phages against dysentery but also against cholera and bubonic plague (Chanishvili 2012). However, the lack of standardized methods including proper controls has resulted in several conflicting studies on the potential of phages in a therapeutic context (Summers 2012). In addition, phages usually have a narrow host range (limited to a few bacterial strains is some cases), which makes their use highly targeted, unlike antibiotics, which have a much broader spectrum (Summers 2012). As indicated in chapter "The Rise and Fall of Antibiotics in Aquaculture," the Second World War was a decisive moment in the discovery and use of antibiotics and thus a breaking point for the use of phages in several countries, with the exception of the Soviet Union and Germany (Cisek et al. 2017).

Nevertheless, even today, phage therapy is still used in several Eastern European countries, such as the Georgia, Poland, and Russia (Abedon et al. 2011). The largest institute dedicated to phage therapy is in Georgia. The Eliava Institute is a historic center founded by the Georgian microbiologist George Eliava in collaboration with Félix d'Hérelle (Sulakvelidze and Alavidze 2001). This unique phage therapy clinic offers highly specialized and personalized treatments for ears, throat, nose, urologic, and gynecologic human infections. With the growing crisis of antibiotic-resistant bacteria, the use of phages in a therapeutic or biocontrol context has been rediscovered (Abedon 2014). This growing interest of phages in a therapeutic context can be illustrated by the increasing number of articles about "phage therapy" that are indexed by Web of Science (Fig. 1).

A recent clinical study has shown excellent success rates for treating chronic otitis caused by *Pseudomonas aeruginosa* with phages (Wright et al. 2009). In addition, a European initiative, PhagoBurn (http://www.phagoburn.eu), is currently investigating phage therapy against *Escherichia coli* and *P. aeruginosa*. From a commercial point of view, several companies are interested in a therapeutic application of phages, and some products based on these viruses are already approved and marketed in Western countries (Sarhan and Azzazy 2015). For example, in Canada, it is possible to use the product AgriPhage-CMM, which contains phages against *Clavibacter michiganensis* subsp. *michiganensis*, the etiological agent causing canker of tomatoes. Phage products are also registered in the USA for food

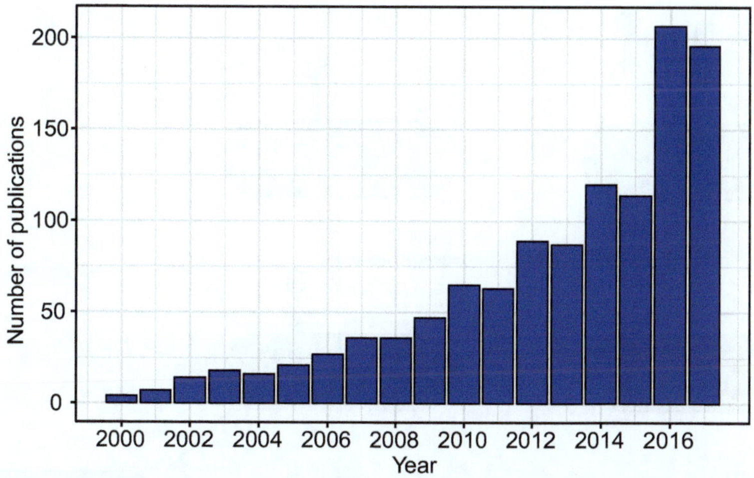

Fig. 1 Occurrence of the term "phage therapy" in Web of Science-indexed publications (ISI) (by choosing "All Databases") on June 3, 2018

applications (Bai et al. 2016), including among others ListShield (against *Listeria monocytogenes*), EcoShield (against *E. coli* O157: H7), and SalmoFresh (against *Salmonella enterica*).

2 The Biology of Phages

Similarly to the majority of antibiotic molecules (before some artificial chemical modifications), phages are naturally found in the environment, particularly in bacteria-colonized ecosystems. Phages can be found at a titer of up to 2.5×10^8 viral particles per milliliter of water in natural environments (Bergh et al. 1989). In addition, it is now well known that phages play important ecological roles in controlling bacterial populations and participating in the regulation of several biogeochemical cycles (Bratbak et al. 1990; Proctor and Fuhrman 1990; Suttle 2007; Sime-Ngando 2014).

Inside the bacterial host cells, a phage can remain in at least four forms leading to different evolutionary strategies: as a replicating virus during the lytic cycle, as an unstable carrier state termed pseudolysogeny, as a prophage with complete genome during the lysogeny state, or as a defective cryptic prophage (Golais et al. 2013). To simplify, we often talk about the lytic cycle in opposition to the lysogenic state (Fig. 2). A lytic cycle comprises five major steps (Sulakvelidze and Alavidze 2001; Drulis-Kawa et al. 2012): (I) adsorption of the phage on the surface of the bacterium, (II) injection of the phage genetic material in the host bacterium, (III) phage genome replication and host cell machinery redirection for viral particle production,

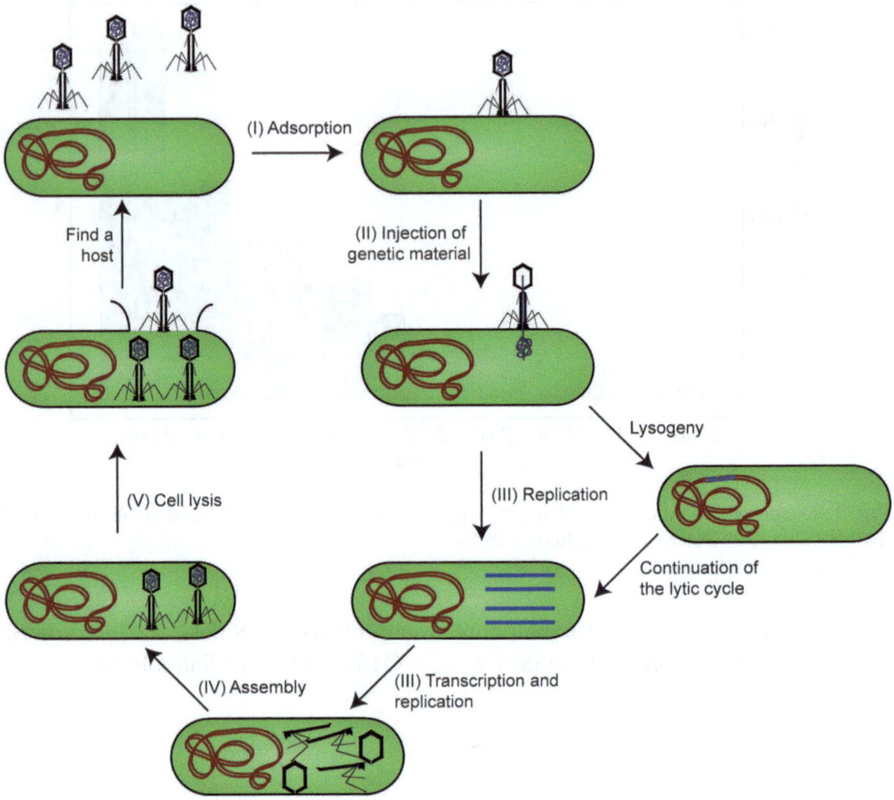

Fig. 2 Schematization of the stages of lytic and lysogenic cycles

(IV) phage assembly, and (V) host cell lysis. For a temperate phage, steps I and II occur as for a lytic phage; however, during lysogeny the phage DNA is inserted into the bacterial chromosome (Sulakvelidze and Alavidze 2001). At this stage, the phage DNA is replicated along with the bacterial DNA and is called a prophage. Any given stress can cause excision of the prophage, allowing it to continue its cycle in steps III, IV, and V as for a lytic phage. It should be noted that some prophages, including those found in the bacteria *Leptospira*, can replicate their DNA in an extrachromosomal manner similar to that of plasmids (Girons et al. 2000; Zhu et al. 2015).

In a therapeutic context, strictly lytic phages are favored because the only outcome is killing the phage-infected bacterium. Indeed, temperate phages can cause some undesirable effects. During the lysogenic cycle, phage DNA becomes an integral part of the bacterial genome and can therefore make its genes usable by the bacterium (De Paepe et al. 2014). Several bacterial genomes are known to contain prophages, and in some cases, the percentage of phage DNA can reach more than 20% of the total genome (Hatfull and Hendrix 2011; Casjens 2003). Prophages are recognized for potentially conferring certain characteristics such as causing an increased in the fitness of the bacterium, protecting against phage infections, as well as enhancing virulence through lysogenic conversion factors (Brüssow et al. 2004). The bacterium *E. coli* O157:H7 is a known example of bacteria with toxin genes (Shiga toxins) from a prophage (Plunkett et al. 1999).

While strictly lytic phages remain by far the favorite option for eradication purposes, recent works expand the horizon of phagotherapy. Indeed, some groups have begun to explore the possibility of using temperate phages naturally present in bacteria or completely engineered (Monteiro et al. 2018) as alternative options. The temperate phage therapy may be useful in some cases where it is hard to isolate lytic phages against some specific pathogenic bacteria. For example, there is a lack of virulent phages that target *Clostridium difficile*, an anaerobic pathogen (Hargreaves and Clokie 2014). To overcome this problem, the combination of temperate and lytic phages was used. The cocktail caused the complete lysis of *C. difficile* in vitro and prevented the appearance of lysogens (Nale et al. 2016).

There are also other biological aspects to consider when using phages in a therapeutic context. For example, the host range of the lytic phage and the ability of the targeted bacteria to evolve a defense mechanism are crucial parameters. Phages tend to be specific to one or a few bacterial species or even a few strains (Hyman and Abedon 2010). This can be problematic in a therapeutic context since it requires knowing precisely the bacterium causing the infection, which can still be challenging. However, it is often suggested that a mixture of phages, in the form of a cocktail, could increase the bacterial host range (Chan and Abedon 2012; Chan et al. 2013). For example, a metagenomic method has revealed that the Intesti-bacteriophage cocktail, initially developed by d'Hérelle and now produced by the Eliava Institute, contains about 23 phages, allowing to target a large number of different bacteria (Zschach et al. 2015).

Bacteria can protect themselves against phage infection by several mechanisms. It is possible to categorize these mechanisms in various classical groups: the inhibition of phage adsorption, the blocking of the entry of phage genetic material, the degradation of phage nucleic acids (CRISPR-Cas and restriction-modification systems), and cell death, the latter known as bacterial abortive infection (Abi) systems (Labrie et al. 2010). However, a plethora of additional and novel phage defense mechanisms are still been identified (Doron et al. 2018; Kronheim et al. 2018). Phages, unlike antibiotics, are dynamic biological entities that can evolve to counteract bacterial protection mechanisms (Samson et al. 2013). Again, it is proposed that the use of a cocktail of phages would be able to decrease the level of resistance to these phages (Lu and Koeris 2011; Abuladze et al. 2008). However, it is recommended to minimize the number of viruses and rigorously test the cocktail to verify that phages have no antagonistic effects and that there is no recombination capability between phage genomes (Mateus et al. 2014; Klumpp and Loessner 2013).

3 Phagotherapy in the Digital Age: What if We Could Make a Custom Phage?

As noted in chapter "The Rise and Fall of Antibiotics in Aquaculture," several antibiotic molecules are naturally produced by various organisms, either to protect themselves or to compete for resources. As reviewed elsewhere (Wright et al. 2014; Nicolaou and Rigol 2018), chemists have used, modified, and recreated these molecules in order to have active, safe, bioavailable, optimized, and economically viable compounds. It is interesting that the chemistry of antibiotics, which arguably began in the laboratory of Paul Ehrlich, was born in the early 1900s, as was the discovery of phages by Twort and d'Hérelle. However, it was not until several important discoveries in molecular biology were made before phage could be modified.

Ironically, phages played a crucial role in the discovery of key enzymes used in molecular biology, such as restriction enzymes (Smith and Welcox 1970) and ligases (Weiss and Richardson 1967). The first genomes sequenced were those of phage MS2 (single-stranded RNA) (Fiers et al. 1976), ΦX174 (single-stranded DNA) (Sanger et al. 1977) and λ (double-stranded DNA) (Sanger et al. 1982). More recently, it has been shown that the CRISPR-Cas system, an adaptive defense of bacteria against exogenous DNAs such as phage genomes and plasmids (Barrangou et al. 2007; Marraffini and Sontheimer 2008), can be used to make targeted double-strand DNA breakage and thus be a powerful tool for genome editing (Jinek et al. 2012). Now, thanks to this system, it is even possible to modify phage genomes very precisely (Martel and Moineau 2014; Lemay et al. 2017).

In addition to modifying an existing genome, it is now possible, thanks to the advances made in molecular biology and bioinformatics, to de novo create synthetic genomes and thus reconstitute organisms. This allowed the team of J. Craig Venter in 2003 to artificially recreate, from oligonucleotides, the genome of the phage ΦX174 (infecting *E. coli*) and to generate infectious virions from this synthetic genome (Smith et al. 2003). Although the creation of synthetic genomes must be ethically accepted (Cho et al. 1999), it is clear that this new discipline has the potential to radically change many facets of science and medicine, including, of course, phage therapy (Kilcher et al. 2018; Kilcher and Loessner 2018).

In this sense, given the advances in genome engineering and synthetic biology, it would be possible to modify/create a phage having features ideally designed for a therapeutic context, such as a larger host range and no lysogenic cycle (Barbu et al. 2016; Brown et al. 2017). For example, it has already been possible to modify the host range of phage T2 (infecting *E. coli*) by changing its long tail fiber genes, *37* and *38*, by those of phage IP008, which naturally has a much larger host spectrum than the one of T2 (Mahichi et al. 2009). Similarly, another study demonstrated that by swapping the same genes of phage T2 with those of phage PP01, infecting specifically the pathogenic *E. coli* O157:H7, the modified phage T2 was also able to infect this bacterium (Yoichi et al. 2005). Since it is possible to more closely control the genes present in phage genomes, it is also realistic to think that synthetic biology will make it easier to satisfy safety requirements for the approval process with the different agencies. Like the antibiotic molecules currently available, it is possible to believe that in the near future, natural, semisynthetic, and synthetic phage products will emerge in the market of antimicrobial compounds.

4 Phage Therapy in Aquaculture

As with all other living things, aquatic animals are subject to diseases. In addition, as discussed in chapter "The Rise and Fall of Antibiotics in Aquaculture," it is increasingly difficult to sustainably treat certain bacterial diseases because of the rise of antibiotic-resistant strains. This is why, in recent years, several studies have investigated the potential of phage therapy in the aquatic environments (Gon Choudhury et al. 2017). For example, phages have been tested to control infections caused by *Vibrio* (vibriosis disease) (Wang et al. 2017; Kalatzis et al. 2018), *Flavobacterium columnare* (columnaris disease) (Prasad et al. 2011; Laanto et al. 2015), *Thalassomonas loyana* (white plague coral disease) (Atad et al. 2012), and *Aeromonas salmonicida* subsp. *salmonicida* (furunculosis disease) (Imbeault et al. 2006; Silva et al. 2016). For a little over a year now, it is possible for fish farmers to obtain the product BAFADOR®, a new bacteriophage-based preparation, to fight two of the most common pathogens responsible for mortality in farmed fish, *Aeromonas hydrophila* and *Pseudomonas fluorescens* (globalaginvesting.com). This preparation can be used both prophylactically to increase the resistance of the fish and eels and therapeutically in case of infection (Schulz et al. 2019).

While these studies all show promising results, the fact remains that there are significant challenges inherent to aquaculture that must be considered. One of these challenges is undoubtedly the method of phage delivery (Gon Choudhury et al. 2017). Unlike humans or large animals, with which it is possible to use the oral or intracutaneous routes, the administration of drugs has additional constraints with fish. Several ways of administration were already been proposed. Among these, there is the immersion of fish in a solution containing phages, the addition of viruses in the feed or even their addition directly in water. Finally, other ways also include the anal intubation or the injection of phages. However, these last two methods require important handling of fish that may be difficult to put in place in a fish-farming context. In addition the phage delivery via the parenteral route, i.e., intraperitoneal injection in fish, against *Flavobacterium* resulted in immediate distribution of phages to the circulation system and organs, but no significant reduction of fish mortality was detected (Castillo et al. 2012). By the oral route, the immersion, or directly in circulation system, the phage treatment is likely to be diluted in water (Christiansen et al. 2014). Although it is possible to add phages to food (Nakai and Park 2002; Park and Nakai 2003; Jun et al. 2013), it is well known that fish compete for it (Cuenco et al. 1985); phage intake would consequently be unequal between individuals, especially for those who have less energy to feed oneself like sick fish.

Some other aspects must be challenged with phage therapy in aquaculture. One of them is the bacterial community that is very variable over seasons and years, showing a higher complexity during warm seasons. On the other hand, it seems that the diversity of pathogenic bacteria showed lower seasonal variation as reported for *Vibrio* genus (Pereira et al. 2011a, b). It was suggested that the spring season is the best time to apply phages and that an annual water monitoring is needed. It is also important to consider that the chemical disinfection use in the farms is often a source of disturbance and variation of the microbial community.

It is crucial to identify the main pathogenic bacteria and to choose the best phage cocktail. This information must be rapidly known to avoid the spreading of the infectious agents. The culture-independent in situ hybridization using specific probes provides an overview of the real proportion of cultivable and non-cultivable pathogenic bacteria (Taylor-Brown et al. 2017), or multiplex PCR approaches can be used to diagnostic in a short time (Nishiki et al. 2018). In fact, an annual follow-up of the bacterial diversity could be a good way to use the phages in preventive treatment instead of curative in aquaculture context.

Finally, the phage therapy in aquaculture could be possible if phages show a good survival time in water system according to the chosen method of administration, have only a moderated impact on the overall bacterial community structure, and provide desired specificity and efficiency for the targeted pathogenic bacteria (Pereira et al. 2011a, b).

5 Toward Phagotherapy to Control Furunculosis

The bacterium *A. salmonicida* subsp. *salmonicida* is the etiologic agent of furunculosis, a worldwide disease affecting salmonids (Dallaire-Dufresne et al. 2014). Historically, the control of furunculosis is through vaccination and antibiotic therapy. While vaccination implies heavy fish handling and high cost, the use of antibiotics was, and still is, the preferred method for treating furunculosis. However, more and more cases of *A. salmonicida* subsp. *salmonicida* strains resistant or even multiresistant to antibiotics are listed (Vincent et al. 2014, 2016a; Trudel et al. 2016; Bartkova et al. 2017). With increasing access to high-throughput sequencing technologies, it is now possible to effectively investigate the determinants of resistance to different antimicrobial compounds (Chan 2016). In the case of *A. salmonicida* subsp. *salmonicida* and as discussed in chapter "The Rise and Fall of Antibiotics in Aquaculture," several recent studies have shown that plasmids are important vectors in the spread of antibiotic resistance genes. A study has shown, for example, that two plasmids, pSN254b and pAB5S9b, can provide resistance to all antibiotics approved by the Veterinary Drugs Directorate (VDD) of Health Canada to treat infected fish (Trudel et al. 2016). It is therefore clear that effective alternatives to antibiotics are needed to control furunculosis.

Several independent studies showed in vitro that phage therapy might be one of the alternatives to antibiotics against furunculosis. Many lytic phages infecting

A. salmonicida subsp. *salmonicida* have been found and characterized (Vincent et al. 2017a). Importantly, some of these phages can infect a large number of bacterial strains without distinction of geographical origin or other parameters. Although it is often considered preferable to have phages with a large host range to facilitate treatment, it is also important to avoid using phages that can lyse non-targeted species or strains. In the case of *A. salmonicida*, this seems to be guaranteed by a biological barrier imposed by the bacterium itself. The strains of the subspecies *salmonicida* are psychrophilic, so they cannot grow at 37 °C (Dallaire-Dufresne et al. 2014). However, several mesophilic strains of *A. salmonicida* have recently been characterized, although there is no subspecies attribution yet (Vincent et al. 2016a, b, 2017b, 2019). This biological dichotomy also reflected in phage sensitivity, where mesophilic strains are resistant to phages infecting psychrophilic ones (including strains of other psychrophilic subspecies of *A. salmonicida*) (Table 1). It is interesting to note that the mesophilic strain A527 is an exception since it is sensitive to phage 44RR2.8t.2, isolated from a strain of subspecies *salmonicida*. This suggests that the phage receptor 44RR2.8t.2 may be present in strain A527, although still unknown.

Another major concern with the use of phages in aquaculture is how these bacterial viruses are maintained under aquaculture conditions. Three phages infecting *A. salmonicida* subsp. *salmonicida* were found to persist in water similar to that found in aquaculture for a long period of time (Fig. 3). While the phages were diluted in water, still a considerable number of virions remained detectable in spite of the absence of host bacteria to ensure their replication. It should also be noted that phages can replicate in the presence of their host, allowing a potential for autodosing and thus minimizing the impact of dilution. Another study has also evaluated the

Table 1 Specificity of *A. salmonicida* ssp. *salmonicida* bacteriophages

Strains	Bacteriophages		
	SW69-9 (HER523)	44RR2.8t.2 (HER98)	65.2 (HER110)
A. salmonicida subsp. *salmonicida* 01-B526[a,p]	***Strong***	*Weak*	*Weak*
A. salmonicida subsp. *salmonicida* 2009-144 K3[b,p]	*Weak*	*Weak*	*Weak*
A. salmonicida subsp. *salmonicida* A449[c,p]	***Strong***	*Weak*	*Weak*
A. salmonicida subsp. *achromogenes* JF3116[c,p]	***Strong***	***Strong***	***Strong***
A. salmonicida subsp. *pectinolytica* 34mel[Tc,p]	**Resistant**	**Resistant**	**Resistant**
A. veronii biovar *sobria* AH-42[c,p]	**Resistant**	**Resistant**	**Resistant**
A. salmonicida Y577[c,m]	**Resistant**	**Resistant**	**Resistant**
A. salmonicida A527[c,m]	***Strong***	**Resistant**	***Strong***
A. hydrophila C-1[c,m]	**Resistant**	**Resistant**	**Resistant**

Bold = bacteria resistant to infection by phages; Italic = weak lytic capacity by phages on bacteria; Bold italic = strong lytic capacity by phages on bacteria
p psychrophilic, *m* mesophilic
[a]Origin from Quebec
[b]Origin from Canada
[c]Origin from Europe

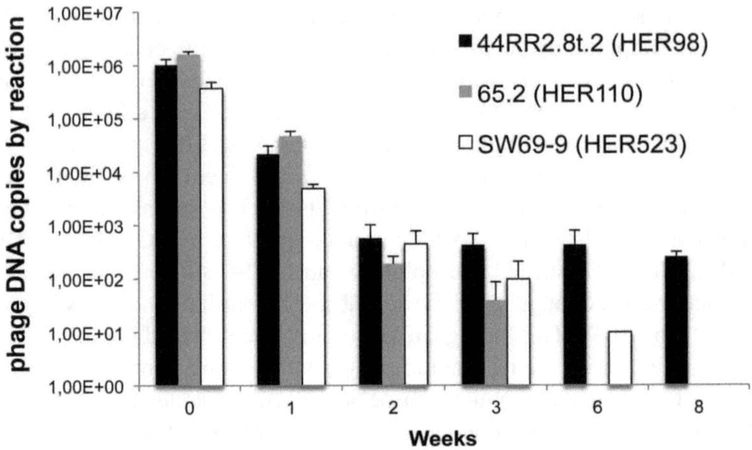

Fig. 3 Detection by qPCR of *A. salmonicida* phage DNA from a phage cocktail in water circulation system in time. Bacteriophages were amplified separately at an initial concentration of 1×10^7 UFP/ml and pooled into the same circulating water system. Time 0 was taken 1 h after the water was completely recircularized. Three water samples were then taken each week. The amount of residual phage DNA was then analyzed by qPCR in triplicate

persistence of *A. salmonicida* subsp. *salmonicida* phages with brook trout by reinfecting fish with doses of *A. salmonicida* subsp. *salmonicida*. After 7 days, the phages remained in aquariums and allowed decrease of the pathogen population (Imbeault et al. 2006).

The emergence of bacteriophage-insensitive mutants (BIM) during phage infection has been reported in some studies in aquaculture context (Hossain et al. 2012; Tan et al. 2015), but the mechanisms of phage resistance are not yet completely understood. The presence of quorum-sensing-regulated phage defense mechanisms depending on population cell density and mutation of the phage receptor seem to be probable strategies to resist phage infection (Hossain et al. 2012). In *A. salmonicida* subsp. *salmonicida*, some phage-resistant bacteria also emerged after phage treatment including successive streak-plating steps (Moreirinha et al. 2018). A significant modification in the expression of intracellular proteins was observed when compared with the phage-sensitive bacteria. These proteins would have molecular function associated in phage replication. This decrease of proteins in the host cell prevented the phage from completing its lytic cycle. Fortunately, phages have also developed strategies to overcome bacterial resistance (Samson et al. 2013) making this constant battle very interesting for therapeutic purpose.

The use of a phage cocktail rather than a single phage for therapeutic purpose has been mentioned in the previous section to avoid bacterial resistance, at least, to decrease it. Although mixing several phages in a same cocktail can make it more effective or synergistic, it can also result in antagonistic effects. A cocktail combining two specific phages against *A. salmonicida* showed significantly higher antimicrobial activity than other cocktails (with three, four, or five phages) and individual phages (Chen et al. 2018).

Fish immune responses are also possible (Khan Mirzaei et al. 2016) if there are bacterial debris present in the phage cocktail (Abedon 2014; Dufour et al. 2016). It would consequently be crucial to develop a protocol for the production and purification of *A. salmonicida* subsp. *salmonicida* phages. Attempts to optimize such general processes have already been made (Bourdin et al. 2014; Lipinski et al. 2016). Finally, the determination of an experimental animal model is of great importance in the study of fish infection to avoid variations and to be close to the reality (Romero et al. 2016).

What about the in vivo trials in aquaculture context against furunculosis? To date, only few publications mention the use of phages against furunculosis (Imbeault et al. 2006; Silva et al. 2016). One of them demonstrated that phage therapy can increase the survival of infected rainbow trout model against an infection with *A. salmonicida* subsp. *salmonicida* (Kim et al. 2015). The intramuscular administration of single phage dose at MOI of 10,000 against *A. salmonicida* subsp. *salmonicida* showed notable protective effects such as increasing the survival rates. For all the smallest MOI, the bacterial growth was markedly retarded up to 12 h after phage inoculation but started to increase gradually after 24 h. Furthermore, no development of fish humoral immunity was shown in this study. These results demonstrated that some phages could be considered as alternative biological control agents against *A. salmonicida* subsp. *salmonicida* infections in rainbow trout, but the required concentration of phages (MOI) is very high. Future works will have to focus on isolating and characterizing phages with high replication rate at lower doses of infection and to test other administration routes.

The early stage of fish growth is particularly important because traditional antibiotic treatments or vaccination are not effective. The implementation of approaches to control furunculosis in juvenile fish is essential to ensure the sustainability of production, and phage therapy could be attractive in this context. After submerging juvenile *Solea senegalensis* in a phage bath for 6 h, the growth of *A. salmonicida* was inhibited. After 72 h, none of the fish had died, while a mortality rate of 36% was reported when fish were exposed only to the pathogen without phages (Silva et al. 2016).

On a less positive note, a 2007 study from the UK on Atlantic salmon and rainbow trout (Verner-Jeffreys et al. 2007) showed that phage administration by the intraperitoneal route or by oral infeed against did not offer protection for fish against *A. salmonicida* subsp. *salmonicida*. Because promising results in vitro do not seem to be always reflected in in vivo challenges (Tsonos et al. 2014), more research studies are clearly needed on the use of phages to treat/prevent fish infections by *A. salmonicida* subsp. *salmonicida*.

6 Keep Going Until Efficient Phagotherapy

With the emergence of multidrug-resistant bacterial strains, we can observe the return in force of the phagotherapy. The natural, specific, and quick way that phages eliminate bacteria suggest the possibility of creating an alternative or complement treatment that is effective and simple against pathogenic bacteria in aquaculture (Dy et al. 2018), as already done in other fields. On the other hand, several parameters have yet to be optimized, such as isolation of hyper-efficient phages accompanied with a well genomic and phenotypic characterization. Understanding the host range and ensuring that the phages will not transfer unwanted genes to the bacterial community is a priority. It will also be of prime importance to increase our understanding about the mechanisms used by pathogenic bacteria to protect themselves against phages. In this sense, it will be crucial to avoid repeating the same mistakes made with antibiotics. Finally, the increase of in vivo experiments could allow us to determine the best route of administration in an aquaculture context and to confirm the feasibility of this approach.

This chapter shows that the commercial use of phages against fish diseases has still some hurdles to clear. However, the fact that phage-based products are already marketed might help to pave the way for more similar products in aquaculture and specifically against furunculosis. It is hoped that the BAFADOR® product offered against *P. fluorescens* and *A. hydrophila* to protect and cure farmed eel is just the beginning.

In parallel, other alternative treatments to antibiotics should be deployed to avoid relying on a single therapeutic strategy. For example, improving the composition and methods of vaccination in aquaculture is a topic of research for many groups (Gudding et al. 1999; Hastein et al. 2005; Kashulin et al. 2017). It has been also demonstrated that phages are more effective when used with antibiotics, regardless of the antibiotic resistance state of the bacterium (Comeau et al. 2007). This phenomenon, called the phage-antibiotic synergy, suggested that other molecules or combined treatments can also be synergistic with phages. For example, carvacrol, an essential oil, combined with pneumococcal phage lysozyme improves their lytic activity (Díez-Martínez et al. 2013). The use of probiotics is also a popular approach (see chapter "Host-Microbiota Interactions and their Importance in Promoting Growth and Resistance to Opportunistic Diseases in Salmonids") in aquaculture (Das et al. 2008; Hai 2015) that could be combined with phagotherapy.

Governmental approval will also likely be needed for these phage-based products. The acceptance of the phage technology and its adoption by the fish farmers and consumers may dictate its commercial success. If fish consumers and producers are willing to accept phage treatments once proven to be effective, reproducible, and safe, then different governmental agencies will be more prone to approve the use of these products.

References

Abedon ST (2014) Phage therapy: eco-physiological pharmacology. Scientifica (Cairo) 2014:581639. https://doi.org/10.1155/2014/581639

Abedon ST, Kuhl SJ, Blasdel BG, Kutter EM (2011) Phage treatment of human infections. Bacteriophage 1:66–85. https://doi.org/10.4161/bact.1.2.15845

Abuladze T et al (2008) Bacteriophages reduce experimental contamination of hard surfaces, tomato, spinach, broccoli, and ground beef by *Escherichia coli* O157:H7. Appl Environ Microbiol 74:6230–6238. https://doi.org/10.1128/AEM.01465-08

Atad I, Zvuloni A, Loya Y, Rosenberg E (2012) Phage therapy of the white plague-like disease of *Favia favus* in the Red Sea. Coral Reefs 31:665–670. https://doi.org/10.1007/s00338-012-0900-5

Bai J, Kim Y-T, Ryu S, Lee J-H (2016) Biocontrol and rapid detection of food-borne pathogens using bacteriophages and endolysins. Front Microbiol 7:474. https://doi.org/10.3389/fmicb.2016.00474

Barbu EM, Cady KC, Hubby B (2016) Phage therapy in the era of synthetic biology. Cold Spring Harb Perspect Biol 8:a023879. https://doi.org/10.1101/cshperspect.a023879

Barrangou R et al (2007) CRISPR provides acquired resistance against viruses in prokaryotes. Science 315:1709–1712. https://doi.org/10.1126/science.1138140

Bartkova S, Leekitcharoenphon P, Aarestrup FM, Dalsgaard I (2017) Epidemiology of Danish *Aeromonas salmonicida* subsp. *salmonicida* in fish farms using whole genome sequencing. Front Microbiol 8:2411. https://doi.org/10.3389/fmicb.2017.02411

Bergh O, Børsheim KY, Bratbak G, Heldal M (1989) High abundance of viruses found in aquatic environments. Nature 340:467–468. https://doi.org/10.1038/340467a0

Bourdin G et al (2014) Amplification and purification of T4-Like *Escherichia coli* phages for phage therapy: from laboratory to pilot scale. Appl Environ Microbiol 80:1469–1476. https://doi.org/10.1128/AEM.03357-13

Bratbak G, Heldal M, Norland S, Thingstad TF (1990) Viruses as partners in spring bloom microbial trophodynamics. Appl Environ Microbiol 56:1400–1405

Brown R, Lengeling A, Wang B (2017) Phage engineering: how advances in molecular biology and synthetic biology are being utilized to enhance the therapeutic potential of bacteriophages. Quant Biol 5:42–54. https://doi.org/10.1007/s40484-017-0094-5

Brüssow H, Canchaya C, Hardt W-D (2004) Phages and the evolution of bacterial pathogens: from genomic rearrangements to lysogenic conversion. Microbiol Mol Biol Rev 68:560–602. https://doi.org/10.1128/MMBR.68.3.560-602.2004

Casjens S (2003) Prophages and bacterial genomics: what have we learned so far? Mol Microbiol 49:277–300. https://doi.org/10.1046/j.1365-2958.2003.03580.x

Castillo D et al (2012) Diversity of *Flavobacterium psychrophilum* and the potential use of its phages for protection against bacterial cold water disease in salmonids. J Fish Dis 35:193–201. https://doi.org/10.1111/j.1365-2761.2011.01336.x

Chan KG (2016) Whole-genome sequencing in the prediction of antimicrobial resistance. Expert Rev Anti-Infect Ther 14:617–619. https://doi.org/10.1080/14787210.2016.1193005

Chan BK, Abedon ST (2012) Phage therapy pharmacology. Phage cocktails. Adv Appl Microbiol 78:1–23. https://doi.org/10.1016/B978-0-12-394805-2.00001-4

Chan BK, Abedon ST, Loc-Carrillo C (2013) Phage cocktails and the future of phage therapy. Future Microbiol 8:769–783

Chanishvili N (2012) Chapter 1—Phage therapy—history from Twort and d'Herelle through soviet experience to current approaches. In: Łobocka M, Szybalski WT (eds) Bacteriophages, part B, Advances in virus research, vol 83. Academic, Amsterdam, pp 3–40. https://doi.org/10.1016/B978-0-12-394438-2.00001-3

Chen L et al (2018) In vitro design and evaluation of phage cocktails against *Aeromonas salmonicida*. Front Microbiol 9:1476. https://doi.org/10.3389/fmicb.2018.01476

Cho MK, Magnus D, Caplan AL, McGee D (1999) Ethical considerations in synthesizing a minimal genome. Science 286:2087–2090. https://doi.org/10.1126/science.286.5447.2087

Christiansen RH, Dalsgaard I, Middelboe M, Lauritsen AH, Madsen L (2014) Detection and quantification of *Flavobacterium psychrophilum*-specific bacteriophages in vivo in rainbow trout upon oral administration: implications for disease control in aquaculture. Appl Environ Microbiol 80:7683–7693. https://doi.org/10.1128/AEM.02386-14

Cisek AA, Dąbrowska I, Gregorczyk KP, Wyżewski Z (2017) Phage therapy in bacterial infections treatment: one hundred years after the discovery of bacteriophages. Curr Microbiol 74:277–283. https://doi.org/10.1007/s00284-016-1166-x

Comeau AM, Tétart F, Trojet SN, Prère MF, Krisch HM (2007) Phage-antibiotic synergy (PAS): β-lactam and quinolone antibiotics stimulate virulent phage growth. PLoS One 2:e799. https://doi.org/10.1371/journal.pone.0000799

Cuenco ML, Stickney RR, Grant WE (1985) Fish bioenergetics and growth in aquaculture ponds: III. Effects of intraspecific competition, stocking rate, stocking size and feeding rate on fish productivity. Ecol Model 28:73–95. https://doi.org/10.1016/0304-3800(85)90014-6

d'Herelle F (1917) Sur un microbe invisible antagoniste des bacilles dysentériques. C R Acad Sci 165:373–375

Dallaire-Dufresne S, Tanaka KH, Trudel MV, Lafaille A, Charette SJ (2014) Virulence, genomic features, and plasticity of *Aeromonas salmonicida* subsp. *salmonicida*, the causative agent of fish furunculosis. Vet Microbiol 169:1–7. https://doi.org/10.1016/j.vetmic.2013.06.025

Das S, Ward LR, Burke C (2008) Prospects of using marine actinobacteria as probiotics in aquaculture. Appl Microbiol Biotechnol 81:419–429. https://doi.org/10.1007/s00253-008-1731-8

De Paepe M et al (2014) Temperate phages acquire DNA from defective prophages by relaxed homologous recombination: the role of Rad52-Like recombinases. PLoS Genet 10:e1004181. https://doi.org/10.1371/journal.pgen.1004181

Díez-Martínez R et al (2013) Improving the lethal effect of Cpl-7, a pneumococcal phage lysozyme with broad bactericidal activity, by inverting the net charge of its cell wall-binding module. Antimicrob Agents Chemother 57:5355–5365. https://doi.org/10.1128/AAC.01372-13

Doron S et al (2018) Systematic discovery of antiphage defense systems in the microbial pangenome. Science 359:eaar4120. https://doi.org/10.1126/science.aar4120

Drulis-Kawa Z, Majkowska-Skrobek G, Maciejewska B, Delattre A-S, Lavigne R (2012) Learning from bacteriophages—advantages and limitations of phage and phage-encoded protein applications. Curr Protein Pept Sci 13:699–722

Dufour N, Henry M, Ricard J-D, Debarbieux L (2016) Commentary: morphologically distinct *Escherichia coli* bacteriophages differ in their efficacy and ability to stimulate cytokine release in vitro. Front Microbiol 7:1029

Dy RL, Rigano LA, Fineran PC (2018) Phage-based biocontrol strategies and their application in agriculture and aquaculture. Biochem Soc Trans 46:1605–1613. https://doi.org/10.1042/BST20180178

Fiers W et al (1976) Complete nucleotide sequence of bacteriophage MS2 RNA: primary and secondary structure of the replicase gene. Nature 260:500–507. https://doi.org/10.1038/260500a0

Girons IS et al (2000) The LE1 bacteriophage replicates as a plasmid within *Leptospira biflexa*: construction of an *L. biflexa-Escherichia coli* shuttle vector. J Bacteriol 182:5700–5705. https://doi.org/10.1128/JB.182.20.5700-5705.2000

Golais F, Hollý J, Vítkovská J (2013) Coevolution of bacteria and their viruses. Folia Microbiol (Praha) 58:177–186. https://doi.org/10.1007/s12223-012-0195-5

Gon Choudhury T, Tharabenahalli Nagaraju V, Gita S, Paria A, Parhi J (2017) Advances in bacteriophage research for bacterial disease control in aquaculture. Rev Fish Sci Aquac 25:113–125. https://doi.org/10.1080/23308249.2016.1241977

Gudding R, Lillehaug A, Evensen (1999) Recent developments in fish vaccinology. Vet Immunol Immunopathol 72:203–212. https://doi.org/10.1016/S0165-2427(99)00133-6

Hai NV (2015) The use of probiotics in aquaculture. J Appl Microbiol 119:917–935. https://doi.org/10.1111/jam.12886

Hargreaves KR, Clokie MRJ (2014) *Clostridium difficile* phages: still difficult? Front Microbiol 5. https://doi.org/10.3389/fmicb.2014.00184

Hastein T, Gudding R, Evensen O (2005) Bacterial vaccines for fish—an update of the current situation worldwide. Dev Biol 121:55–74

Hatfull GF, Hendrix RW (2011) Bacteriophages and their genomes. Curr Opin Virol 1:298–303. https://doi.org/10.1016/j.coviro.2011.06.009

Hossain MJ, Rahman KS, Terhune JS, Liles MR (2012) An outer membrane porin protein modulates phage susceptibility in *Edwardsiella ictaluri*. Microbiology 158:474–487. https://doi.org/10.1099/mic.0.054866-0

Hyman P, Abedon ST (2010) Bacteriophage host range and bacterial resistance. Adv Appl Microbiol 70:217–248. https://doi.org/10.1016/S0065-2164(10)70007-1

Imbeault S, Parent S, Lagacé M, Uhland CF, Blais JF (2006) Using bacteriophages to prevent furunculosis caused by *Aeromonas salmonicida* in farmed brook trout. J Aquat Anim Health 18:203–214. https://doi.org/10.1577/H06-019.1

Jinek M et al (2012) A programmable dual-RNA-guided DNA endonuclease in adaptive bacterial immunity. Science 337:816–821. https://doi.org/10.1126/science.1225829

Jun JW et al (2013) Protective effects of the *Aeromonas* phages pAh1-C and pAh6-C against mass mortality of the cyprinid loach (*Misgurnus anguillicaudatus*) caused by *Aeromonas hydrophila*. Aquaculture 416–417:289–295. https://doi.org/10.1016/j.aquaculture.2013.09.045

Kalatzis GP, Castillo D, Katharios P, Middelboe M (2018) Bacteriophage interactions with marine pathogenic vibrios: implications for phage therapy. Antibiotics 7:E15. https://doi.org/10.3390/antibiotics7010015

Kashulin A, Seredkina N, Sørum H (2017) Cold-water vibriosis. The current status of knowledge. J Fish Dis 40:119–126. https://doi.org/10.1111/jfd.12465

Khan Mirzaei M et al (2016) Morphologically distinct *Escherichia coli* bacteriophages differ in their efficacy and ability to stimulate cytokine release in vitro. Front Microbiol 7:437

Kilcher S, Loessner MJ (2018) Engineering bacteriophages as versatile biologics. Trends Microbiol. https://doi.org/10.1016/j.tim.2018.09.006

Kilcher S, Studer P, Muessner C, Klumpp J, Loessner MJ (2018) Cross-genus rebooting of custom-made, synthetic bacteriophage genomes in L-form bacteria. Proc Natl Acad Sci 115:567–572. https://doi.org/10.1073/pnas.1714658115

Kim JH et al (2015) Biological control of *Aeromonas salmonicida* subsp. *salmonicida* infection in rainbow trout (*Oncorhynchus mykiss*) using *Aeromonas* phage PAS-1. Transbound Emerg Dis 62:81–86. https://doi.org/10.1111/tbed.12088

Klumpp J, Loessner M (2013) New research on bacteriophages and food safety. In: Sofos J (ed) Advances in microbial food safety. Woodhead Publishing Series in Food Science, Technology and Nutrition Elsevier Science, Cambridge, p 321

Kronheim S et al (2018) A chemical defence against phage infection. Nature 564:283–286. https://doi.org/10.1038/s41586-018-0767-x

Laanto E, Bamford JKH, Ravantti JJ, Sundberg LR (2015) The use of phage FCL-2 as an alternative to chemotherapy against columnaris disease in aquaculture. Front Microbiol 6:829. https://doi.org/10.3389/fmicb.2015.00829

Labrie SJ, Samson JE, Moineau S (2010) Bacteriophage resistance mechanisms. Nat Rev Microbiol 8:317–327. https://doi.org/10.1038/nrmicro2315

Lemay ML, Tremblay DM, Moineau S (2017) Genome engineering of virulent lactococcal phages using CRISPR-Cas9. ACS Synth Biol 6:1351–1358. https://doi.org/10.1021/acssynbio.6b00388

Lipinski T, Gamian A, Zuziak E, Korzeniowska-Kowal A, Gorski A (2016) Purified bacteriophage, its preparation and application. http://www.google.com/patents/US9255251

Lu TK, Koeris MS (2011) The next generation of bacteriophage therapy. Curr Opin Microbiol 14:524–531. https://doi.org/10.1016/j.mib.2011.07.028

Mahichi F, Synnott AJ, Yamamichi K, Osada T, Tanji Y (2009) Site-specific recombination of T2 phage using IP008 long tail fiber genes provides a targeted method for expanding host range while retaining lytic activity. FEMS Microbiol Lett 295:211–217. https://doi.org/10.1111/j.1574-6968.2009.01588.x

Marraffini LA, Sontheimer EJ (2008) CRISPR interference limits horizontal gene transfer in *staphylococci* by targeting DNA. Science 322:1843–1845. 322/5909/1843 [pii]. https://doi.org/10.1126/science.1165771

Martel B, Moineau S (2014) CRISPR-Cas: an efficient tool for genome engineering of virulent bacteriophages. Nucleic Acids Res 42:9504–9513. https://doi.org/10.1093/nar/gku628

Mateus L et al (2014) Efficiency of phage cocktails in the inactivation of *Vibrio* in aquaculture. Aquaculture 424–425:167–173. https://doi.org/10.1016/j.aquaculture.2014.01.001

McKinley EB (1923) The bacteriophage in the treatment of infections. Arch Intern Med 32:899–910. https://doi.org/10.1001/archinte.1923.00110240092005

Monteiro R, Pires DP, Costa AR, Azeredo J (2018) Phage therapy: going temperate? Trends Microbiol. https://doi.org/10.1016/J.TIM.2018.10.008

Moreirinha C et al (2018) Protein expression modifications in phage-resistant mutants of *Aeromonas salmonicida* after AS-A phage treatment. Antibiotics 7:21. https://doi.org/10.3390/antibiotics7010021

Nakai T, Park SC (2002) Bacteriophage therapy of infectious diseases in aquaculture. Res Microbiol 153:13–18. https://doi.org/10.1016/S0923-2508(01)01280-3

Nale JY et al (2016) Bacteriophage combinations significantly reduce *Clostridium difficile* growth in vitro and proliferation in vivo. Antimicrob Agents Chemother 60:968–981. https://doi.org/10.1128/AAC.01774-15

Nicolaou KC, Rigol S (2018) A brief history of antibiotics and select advances in their synthesis. J Antibiot (Tokyo) 71:153–184. https://doi.org/10.1038/ja.2017.62

Nishiki I, Minami T, Murakami A, Hoai TD, Fujiwara A (2018) Multilocus sequence analysis of Vibrionaceae isolated from farmed amberjack and the development of a multiplex PCR assay for the detection of pathogenic species. J Fish Dis 41:1295–1301. https://doi.org/10.1111/jfd.12823

Park SC, Nakai T (2003) Bacteriophage control of *Pseudomonas plecoglossicida* infection in ayu *Plecoglossus altivelis*. Dis Aquat Org 53:33–39. https://doi.org/10.3354/dao053033

Pereira C et al (2011a) Bacteriophages with potential for inactivation of fish pathogenic bacteria: survival, host specificity and effect on bacterial community structure. Mar Drugs 9:2236–2255. https://doi.org/10.3390/md9112236

Pereira C et al (2011b) Evaluating seasonal dynamics of bacterial communities in marine fish aquaculture: a preliminary study before applying phage therapy. J Environ Monit 13:1053–1058. https://doi.org/10.1039/c0em00434k

Plunkett G, Rose DJ, Durfee TJ, Blattner FR (1999) Sequence of Shiga toxin 2 phage 933W from *Escherichia coli* O157:H7: Shiga toxin as a phage late-gene product. J Bacteriol 181:1767–1778

Prasad Y, Arpana KD, Sharma AK (2011) Lytic bacteriophages specific to *Flavobacterium columnare* rescue catfish, *Clarias batrachus* (Linn.) from columnaris disease. J Environ Biol 32:161–168

Proctor LM, Fuhrman JA (1990) Viral mortality of marine bacteria and cyanobacteria. Nature 343:60–62. https://doi.org/10.1038/343060a0

Romero A et al (2016) The animal model determines the results of *Aeromonas* virulence factors. Front Microbiol 7:1574. https://doi.org/10.3389/fmicb.2016.01574

Samson JE, Magadán AH, Sabri M, Moineau S (2013) Revenge of the phages: defeating bacterial defences. Nat Rev Microbiol 11:675–687. https://doi.org/10.1038/nrmicro3096

Sanger F et al (1977) Nucleotide sequence of bacteriophage phi X174 DNA. Nature 265:687–695

Sanger F, Coulson AR, Hong GF, Hill DF, Petersen GB (1982) Nucleotide sequence of bacteriophage lambda DNA. J Mol Biol 162:729–773. https://doi.org/10.1016/0022-2836(82)90546-0

Sarhan WA, Azzazy HME (2015) Phage approved in food, why not as a therapeutic? Expert Rev Anti-Infect Ther 13:91–101. https://doi.org/10.1586/14787210.2015.990383

Schulz P, Robak S, Dastych J, Siwicki AK (2019) Influence of bacteriophages cocktail on European eel (*Anguilla anguilla*) immunity and survival after experimental challenge. Fish Shellfish Immunol 84:28–37. https://doi.org/10.1016/j.fsi.2018.09.056

Silva YJ et al (2016) Biological control of *Aeromonas salmonicida* infection in juvenile Senegalese sole (*Solea senegalensis*) with Phage AS-A. Aquaculture 450:225–233. https://doi.org/10.1016/j.aquaculture.2015.07.025

Sime-Ngando T (2014) Environmental bacteriophages: viruses of microbes in aquatic ecosystems. Front Microbiol 5:1–14. https://doi.org/10.3389/fmicb.2014.00355

Smith HO, Welcox KW (1970) A restriction enzyme from *Hemophilus influenzae*. I. Purification and general properties. J Mol Biol 51:379–391. https://doi.org/10.1016/0022-2836(70)90149-X

Smith HO, Hutchison CA, Pfannkoch C, Venter JC (2003) Generating a synthetic genome by whole genome assembly: X174 bacteriophage from synthetic oligonucleotides. Proc Natl Acad Sci 100:15440–15445. https://doi.org/10.1073/pnas.2237126100

Sulakvelidze A, Alavidze Z (2001) Bacteriophage therapy. Antimicrob Agents Chemother 45:649–659. https://doi.org/10.1128/AAC.45.3.649

Summers WC (2012) The strange history of phage therapy. Bacteriophage 2:130–133. https://doi.org/10.4161/bact.20757

Suttle CA (2007) Marine viruses—major players in the global ecosystem. Nat Rev Microbiol 5:801–812. https://doi.org/10.1038/nrmicro1750

Tan D, Svenningsen SL, Middelboe M (2015) Quorum sensing determines the choice of antiphage defense strategy in *Vibrio anguillarum*. MBio 6:e00627. https://doi.org/10.1128/mBio.00627-15

Taylor-Brown A et al (2017) Culture-independent genomics of a novel chlamydial pathogen of fish provides new insight into host-specific adaptations utilized by these intracellular bacteria. Environ Microbiol 19:1899–1913. https://doi.org/10.1111/1462-2920.13694

Trudel MV et al (2016) Diversity of antibiotic-resistance genes in Canadian isolates of *Aeromonas salmonicida subsp. salmonicida*: dominance of pSN254b and discovery of pAsa8. Sci Rep 6. https://doi.org/10.1038/srep35617

Tsonos J et al (2014) Hurdles in bacteriophage therapy: deconstructing the parameters. Vet Microbiol 171:460–469. https://doi.org/10.1016/j.vetmic.2013.11.001

Twort FW (1915) An investigation on the nature of ultra-microscopic viruses. Lancet 186:1241–1243

Verner-Jeffreys DW et al (2007) Furunculosis in Atlantic salmon (*Salmo salar L.*) is not readily controllable by bacteriophage therapy. Aquaculture 270:475–484. https://doi.org/10.1016/j.aquaculture.2007.05.023

Vincent AT et al (2014) Detection of variants of the pRAS3, pAB5S9, and pSN254 plasmids in *Aeromonas salmonicida subsp. salmonicida*: multidrug-resistance, interspecies exchanges, and plasmid reshaping. Antimicrob Agents Chemother 58:7367–7374. https://doi.org/10.1128/AAC.03730-14

Vincent AT, Emond-Rheault J-G et al (2016a) Antibiotic resistance due to an unusual ColE1-type replicon plasmid in *Aeromonas salmonicida*. Microbiology 162:942–953. https://doi.org/10.1099/mic.0.000286

Vincent AT, Trudel MV et al (2016b) Increasing genomic diversity and evidence of constrained lifestyle evolution due to insertion sequences in *Aeromonas salmonicida*. BMC Genomics 17:44. https://doi.org/10.1186/s12864-016-2381-3

Vincent AT, Paquet VE, Bernatchez A, Tremblay DM, Moineau S, Charette SJ (2017a) Characterization and diversity of phages infecting *Aeromonas salmonicida* subsp. *salmonicida*. Sci Rep 7(1). https://doi.org/10.1038/s41598-017-07401-7

Vincent AT, Rouleau FD, Moineau S, Charette SJ (2017b) Study of mesophilic *Aeromonas salmonicida* A527 strain sheds light on the species' lifestyles and taxonomic dilemma. FEMS Microbiol Lett 364:fnx239. https://doi.org/10.1093/femsle/fnx239

Vincent AT et al (2019) Investigation of the virulence and genomics of *Aeromonas salmonicida* strains isolated from human patients. Infect Genet Evol 68:1–9. https://doi.org/10.1016/J.MEEGID.2018.11.019

Wang Y et al (2017) Bacteriophage therapy for the control of *Vibrio harveyi* in greenlip abalone (*Haliotis laevigata*). Aquaculture 473:251–258. https://doi.org/10.1016/j.aquaculture.2017.01.003

Weiss B, Richardson CC (1967) Enzymatic breakage and joining of deoxyribonucleic acid, I. Repair of single-strand breaks in DNA by an enzyme system from *Escherichia coli* infected with T4 bacteriophage. Proc Natl Acad Sci U S A 57:1021–1028. https://doi.org/10.1073/pnas.57.4.1021

Wright A, Hawkins CH, Änggård EE, Harper DR (2009) A controlled clinical trial of a therapeutic bacteriophage preparation in chronic otitis due to antibiotic-resistant *Pseudomonas aeruginosa*; a preliminary report of efficacy. Clin Otolaryngol 34:349–357. https://doi.org/10.1111/j.1749-4486.2009.01973.x

Wright PM, Seiple IB, Myers AG (2014) The evolving role of chemical synthesis in antibacterial drug discovery. Angew Chem Int Ed Engl 53:8840–8869. https://doi.org/10.1002/anie.201310843

Yoichi M, Abe M, Miyanaga K, Unno H, Tanji Y (2005) Alteration of tail fiber protein *gp38* enables T2 phage to infect *Escherichia coli* O157:H7. J Biotechnol 115:101–107. https://doi.org/10.1016/j.jbiotec.2004.08.003

Zhu W et al (2015) Identification of three extra-chromosomal replicons in *Leptospira pathogenic* strain and development of new shuttle vectors. BMC Genomics 16:90. https://doi.org/10.1186/s12864-015-1321-y

Zschach H et al (2015) What can we learn from a metagenomic analysis of a Georgian bacteriophage cocktail? Viruses 7:6570–6589. https://doi.org/10.3390/v7122958

Controlling Factors for Community Assembly in Developing Cod Larvae (*Gadus morhua*)

Ingrid Bakke, Kari Johanne Kihle Attramadal, Ragnhild Inderberg Vestrum, and Olav Vadstein

Abstract Marine fish larvae are characterized by immature immune and digestive systems at hatching. Therefore, they are particularly vulnerable to detrimental interactions with microbes. In this chapter, we review studies performed in our group on the microbiota of cod larvae. The overall aim of these studies was to clarify which factors affected the composition of the larval microbiota in the early developmental stages and thereby be able to identify possible rearing strategies to promote positive microbe–larvae interactions in aquaculture rearing systems. Through careful experimental designs, we managed to separate the effects of the microbiota in rearing water and diet and were surprised to find that the water microbiota seemed to be a more important determinant for the composition of the larval microbiota than the feed. The larval microbiota changed over time, and our results indicated that this was due to developmental changes in the gastrointestinal system and that selection in the digestive tract of the host structured the larval microbiota. We further tested the potential for manipulating the microbiota of larvae through introduction of probiotic candidates. This was not successful, as the probionts were only transiently present in the larval microbiota. This observation was independent of the developmental stage of the cod larvae and despite the fact that the strains had been isolated from cod larvae. We suggest that the best strategy for promoting beneficial microbe–larvae interactions in aquaculture systems is to manage the microbial water quality. To obtain this, microbial ecology must be taken into consideration. With proper management and system design, it is possible to maintain stable, K-selected microbial environments in both flow-through and recirculating aquaculture systems (RAS). This will promote beneficial larvae–microbe interactions and improve the viability of the larvae.

I. Bakke (✉) · K. J. K. Attramadal · R. I. Vestrum · O. Vadstein
NTNU Department of Biotechnology and Food Science, Faculty of Natural Sciences and Technology, Norwegian University of Science and Technology, Trondheim, Norway
e-mail: ingrid.bakke@ntnu.no

N. Derome (ed.), *Microbial Communities in Aquaculture Ecosystems*,
https://doi.org/10.1007/978-3-030-16190-3_4

1 Introduction

During the last decade, there has been a renewed interest in host–microbiota interactions, and this has resulted in a revolution of knowledge regarding the many ways microbiota affect the host through both mutualistic and parasitic interactions (McFall-Ngai et al. 2013). However, despite the fact that fish play a central role in the evolution of vertebrates (Vandepoele et al. 2004), the studies are biased with respect to hosts, with a dominance of studies of mammals and relatively few studies of fish (De Schryver and Vadstein 2014).

Fish embryos develop in a bacteria-free environment inside the chorion. Upon hatching, they are exposed to bacteria in the environments, and all external mucosal surfaces are rapidly colonized. After mouth opening, this also includes the gastrointestinal (GI) tract. From studies on germ-free zebrafish, we know that this early colonization is essential for normal development of the gastrointestinal and immune systems. Some responses to the colonization of the gut have been found to be evolutionary conserved and are similar in fish and mammals (Kanther and Rawls 2010). It has also been established that the gut microbes in vertebrates contribute to the host health, e.g., by harvesting energy from food and by synthesizing vitamins/hormones that are not produced by the host itself (Rosenberg and Zilber-Rosenberg 2011). Moreover, in later years, dysbiosis in the human gut microbiota has been associated with an increasing number of diseases but also metabolic, neurologic, respiratory, hepatic, and cardiovascular diseases (Lerner et al. 2017; Wang et al. 2017b). The effects of microbiota dysbiosis in fish have not been studied to the same extent, but there is no reason to assume that fish are less susceptible than mammals when it comes to negative interactions with their microbial colonizers.

The types of microbes that the newly hatched fish are exposed to in the surrounding water vary among environments. In natural oligotrophic waters, the microbes are competing for nutrients and other resources. Such conditions will disfavor opportunistic microbes and select for slow-growing, specialized populations (Vadstein et al. 2018b). The concentration of microbes and the community composition in such environments are probably relatively stable, at least in short-term perspectives.

An aquaculture facility represents a complex microbial ecosystem, and the microbial conditions differ considerably from those in natural environments. Bacteria are introduced to the rearing environment through intake water, feed, and fish excrements (Fig. 1), and there are temporal variations in bacterial loads and community composition, e.g., due to feed load. Due to substantial production of dissolved organic matter, there is also considerable growth within the system (Attramadal et al. 2014; Vadstein et al. 2018a). Furthermore, microbial community composition differs among compartments in the system, such as biofilm communities in biofilters and tank and pipe walls, dispersed communities in the water, and host-associated communities on fish skin, gills, and GI tract. The interactions taking place between dispersed, biofilm, and host-associated communities are generally poorly understood.

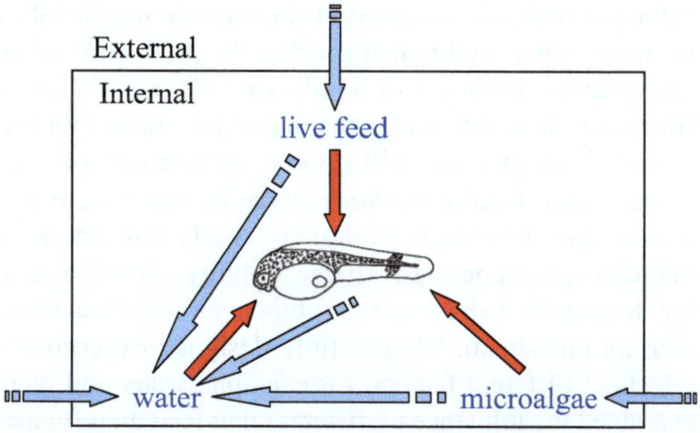

Fig. 1 Important bacterial sources interacting with mucosal surfaces on larval fish (Salvesen 1999)

According to ecological theory, species community composition and diversity are influenced by four basic processes: selection, drift, speciation, and dispersal (Vellend 2010). This also applies to microbial ecosystems (Nemergut et al. 2013). In aquaculture systems, the degree of dispersal limitation depends on the system's design. Drift, or stochastic processes, are probably important in aquaculture, and is likely the reason for the often observed tank-to-tank variation between replicate tanks. Speciation, or diversification, probably has limited influence due to the relative short temporal perspective. The selection acting on the microbial communities in an aquaculture system is made up by a complex set of factors, such as the steps included in the water treatment regime, the hydraulic retention time (both in fish tanks and in the complete system), feeding, competition and interactions within and between communities, etc. (Vadstein et al. 2018a). To control and steer this selection toward favorable microbial conditions is not straightforward, but we have previously shown that recirculating aquaculture systems (RAS) are well suited for managing the microbial rearing water quality to reduce the fraction of opportunistic bacteria (Attramadal et al. 2012b, 2014; Vadstein et al. 2018b).

Even though the number of scientific publications on microbial conditions in aquaculture systems and on fish–microbe interactions has increased considerably during the last years [reviewed in Kelly and Salinas (2017), Wang et al. (2017a, b), and Llewellyn et al. (2014)], many questions remain. How do the microbial communities in the different compartments of the system interact with each other? Is it possible to select for beneficial microbial conditions in aquaculture systems? Which environmental factors are most important in determining the fish microbiota? How do various microbial conditions affect fish health, and what is characteristic to a beneficial fish microbiota? Can we steer the fish microbiota to improve fish health and reduce the risk of disease outbreaks?

Our research group has focused on the interactions between bacteria and marine fish larvae in aquaculture systems, particularly focusing on Atlantic cod (*Gadus morhua*). Both the immune and the intestinal systems of cod are immature at hatching. The pelagic larval stage is characterized by intense development of organs, gut, and immune system, which makes the marine larvae especially vulnerable. Cod larvae are therefore highly susceptible to detrimental microbes, not only to specific pathogens but also to opportunistic bacteria in general (Vadstein et al. 2018a, b). As for many other marine fish species, the production of cod juveniles is a bottleneck, characterized by low and highly variable survival and viability. Previous research strongly indicates that this has microbial causes (Vadstein et al. 2018a, b).

In this chapter, we will review our research on the microbiota associated with Atlantic cod (*Gadus morhua*) larvae in aquaculture systems. A major focus in the reviewed studies has been done to clarify how environmental factors influence the microbiota and consequently the viability of cod larvae and to examine the potential for steering the fish larval microbiota in a beneficial direction, i.e., toward mutualism and not parasitism. We carefully designed experiments to be able to separate the effects of different factors, for example, water and diet. In Bakke et al. (2013), we examined the influence of different live feed diets on the larval microbiota. This was investigated further in Bakke et al. (2015), where we also characterized the development of the larval microbiota with increasing age. We examined the effect of aquaculture systems' water treatment on the rearing water microbiota and larval survival in Attramadal et al. (2014), and in Vestrum et al. (2018b), we investigated how the choice of aquaculture system (flow-through system (FTS) and RAS) affected the rearing water microbiota, the larval microbiota, and eventually the gene expression in the larvae. We also explored the potential for steering the larval microbiota by trying to introduce potentially beneficial bacterial strains (probionts) at different developmental stages (Skjermo et al. 2015). Finally, we summarize these results, draw some conclusions, and suggest implications of our findings for improving larvae–microbe interactions and thus viability of larvae under intensive rearing.

2 Effect of Live Feed Diets on Cod Larval Microbiota

For most marine larvae, first feeding is still dependent on live feed. The live food has its own microbiome (Skjermo and Vadstein 1993; Makridis et al. 2000a), and this microbiome is more abundant than the microbes associated with formulated feed. Relatively few studies have been done on the density and the composition of the microbiota of live feed, but it seems like the microbiota is largely transient and dependent on microbes in the environment (Skjermo and Vadstein 1993; Olsen et al. 2000). This has resulted in the development of methods to control the microbiota of live food (Makridis et al. 2000a; Olsen et al. 2000). However, a recent study on the microbiome of the large copepod *Calanus finmarchicus* from the wild indicates a small set of a core microbiota but with most microbes patchily distributed between individuals (Datta et al. 2018). Both rotifers and *Artemia* are able to ingest bacteria, although with a low efficiency (Vadstein et al. 1993b; Makridis and Vadstein 1999), and therefore they have a constant load of microbes entering their digestive tract. Thus, it should be possible to control the microbiota by changing the composition of microbes in the surrounding water.

Fish larvae also take up microbes from the water, and for marine larvae like Atlantic cod, the uptake of microbes from the water is 100 times higher than the

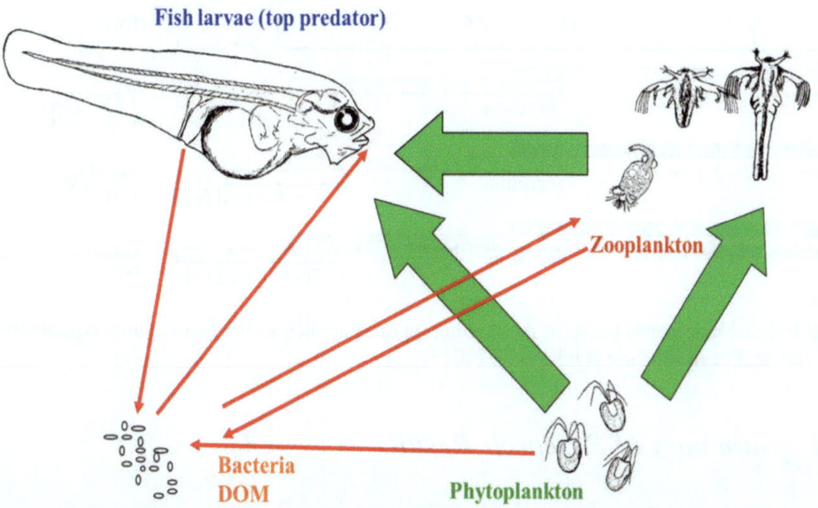

Fig. 2 First feeding of larvae in a complex food web. Arrows indicate direction of effects in the direction of the arrow. Interactions between dissolved organic matter (DOM) and bacteria, and bacteria–bacteria interactions are not indicated. In addition to interactions within the rearing tank, various inputs and losses from the tank also affect the microbe-dominated food web. From Vadstein et al. (2018a)

drinking rate (Reitan et al. 1998). Recent estimations of microbial load to the digestive tract of larvae suggest that microbes from the live food contribute >95% of the total load (Vadstein et al. 2018a). However, it is not evident that this quantitative dominance in the load is materialized in the composition of the gut microbiota as well. Live feed may influence the larval microbiota by two mechanisms: introduction of microbes and selection in the gut due to the biochemical composition of the live food. In addition, the larvae themselves contribute to the selection regime due to the physiochemical environment created in the gut and due to selection setup by the immune system or due to microbe–microbe interactions (Vadstein et al. 2013). Relatively few studies have tried to quantify these various drivers for the composition of the gut microbiota of larvae (Vadstein et al. 2018a).

The first feeding rearing tank of marine larvae can be considered an ecosystem, with a large number of interactions between biota (Fig. 2). Traditionally, the first feeding rearing tank has been considered to be a simple system involving fish, live feed, and microalgae, but including the microbiota the system complexity increases dramatically (Vadstein et al. 2018a). Moreover, in Fig. 2, the microbiota is presented in a simplistic way, as only one functional group, but in reality, it consists of a huge number of species with many possible interactions.

Below we will use data from experiments with Atlantic cod to shed some more light on the importance of live feed (versus water) microbes on the composition of the gut microbiota of marine larvae and try to evaluate the significance of selection in the digestive tract.

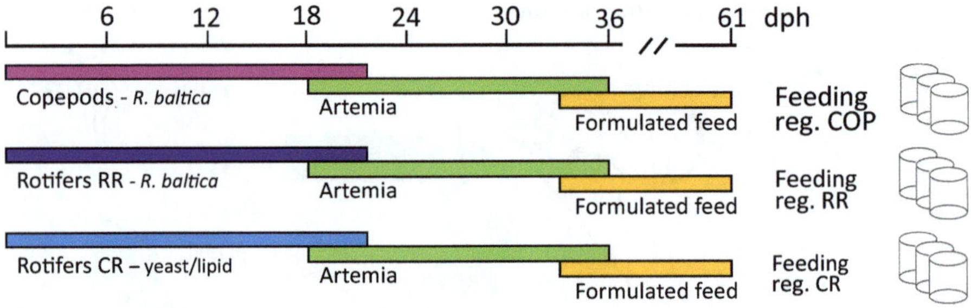

Fig. 3 Schematic presentation of the feeding regimes applied in the first feeding experiment with cod larvae. For details, see Bakke et al. (2013)

2.1 *Summary of Research Results*

To investigate the role of diet on the cod larval microbiota, we designed a first feeding experiment where we used different feeding regimes (Bakke et al. 2013). Three diets were applied from 3 to 22 days post hatching (dph): the copepod *Acartia tonsa* cultivated on microalgae *R. baltica* (diet COP), rotifers (*Brachionus 'Nevada'*) cultivated on *R. baltica* (diet RR), and conventional rotifers (*Brachionus 'Nevada'*) cultivated on yeast/lipids (diet CR). The larvae were then fed with *Artemia franciscana* from 18 to 36 dph and formulated feed from 33 to 61 dph. Figure 3 summarizes the feeding regimes. For each feeding regime, larvae were reared in three replicate 100 liter tanks at a density of 85 larvae per liter.

Bacterial communities associated with cod larvae and live feed were investigated using DGGE analysis of 16S SSU rRNA gene amplicons. We found that distinct bacterial communities were associated with the live feed diets COP, RR, and CR. However, surprisingly, the diet appeared to have little influence on the larval microbiota. Several observations supported this. First, only minor differences in larval microbiota were detected between the COP, RR, and CR rearing tanks. At 17 dph, there were no significant differences in larval microbiota between any of the feeding regimes. Second, despite the change in live feed diet at 18 dph, the larval microbiota was strikingly similar at 17 and 32 dph (Fig. 4). Third, the larval microbiota was generally more similar to the water than the live feed microbiota (Fig. 5). Another interesting finding, also suggesting a potential influence of the water microbiota, was that rearing larvae in replicate tanks (identical rearing regimes) resulted in significant differences in larval microbiota.

In a follow-up study, we analyzed water, live feed, and larval samples for the COP and RR rearing regimes from this first feeding experiment by 16S SSU rRNA gene amplicon pyrosequencing (Bakke et al. 2015). Generally, the results corroborated those obtained by the DGGE analysis. The larval microbiota was similar between feeding regimes and between larval samples from 17 to 32 dph (Fig. 6). Again, the larval microbiota was found to be highly distinct from the live feed microbiota (average Bray–Curtis dissimilarities as high as around 0.95). The

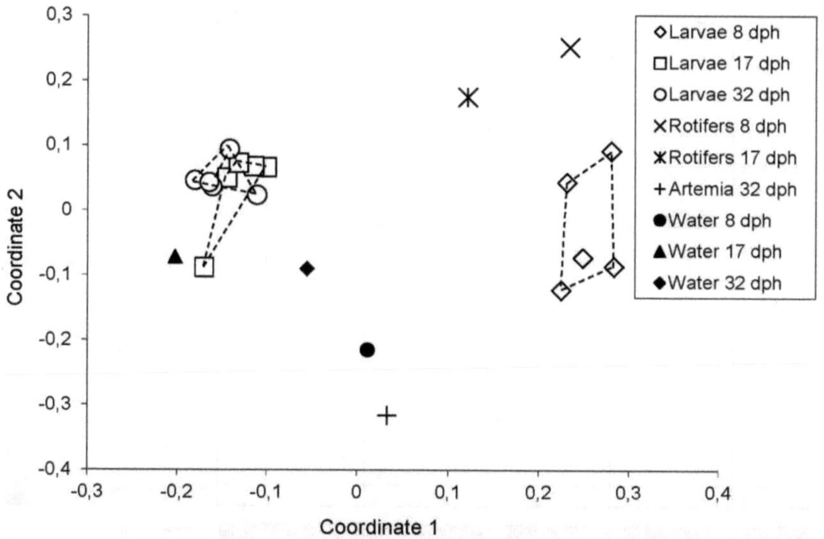

Fig. 4 NMS (nonmetric multidimensional scaling) ordination based on Bray–Curtis similarities for microbiota of live feed, water, and cod larvae in one rearing tank for the RR feeding regime. From Bakke et al. (2013)

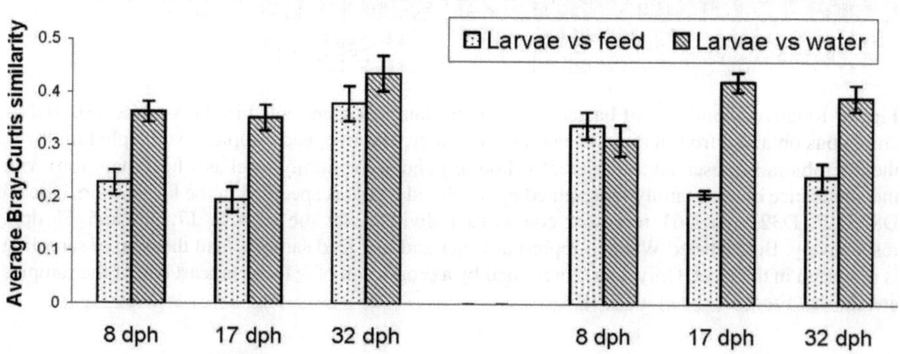

Fig. 5 Average Bray–Curtis similarities for comparisons of cod larvae microbiota with water and live feed microbiota, respectively, among three replicate tanks for the COP and RR rearing systems. Error bars indicate standard errors of the mean. From Bakke et al. (2013)

diversity of the larval microbiota was low at 17 and 32 dph, and an OTU representing *Arcobacter* (*Epsilonproteobacteria*) was highly abundant (Fig. 6). In this study, we suggested that developmental changes in the gastrointestinal system of the larvae could explain the changes in larval microbiota with increasing age and that selection inside the host structured the larval microbiota.

To summarize, the results based on this first feeding experiment with distinct live feed diets suggest that the water microbiota might influence the microbiota associated with developing cod larvae to a greater extent than the live feed. The results

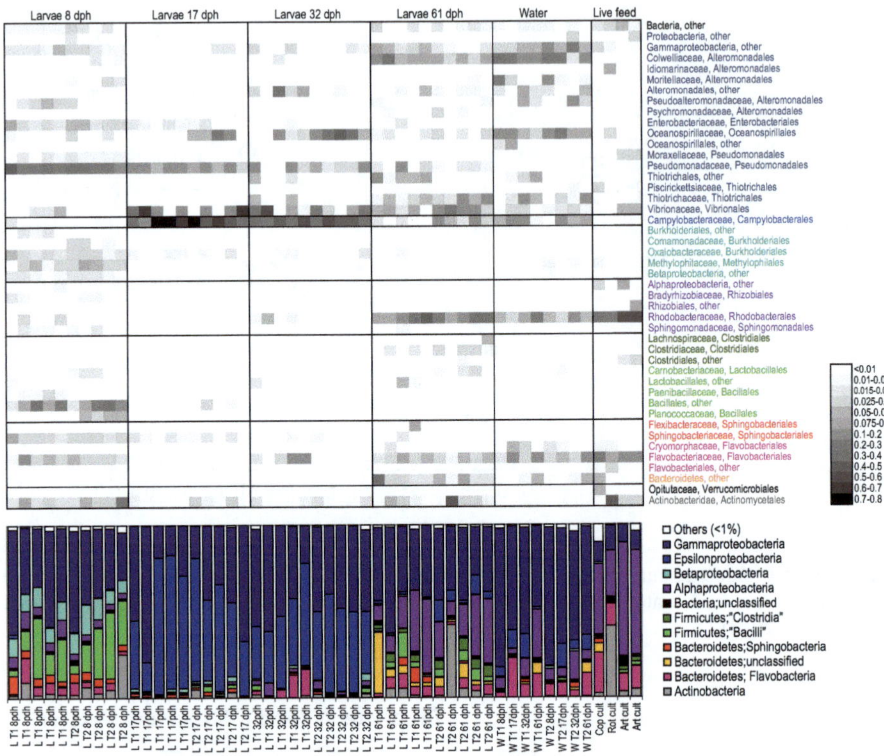

Fig. 6 Relative abundance of bacterial phyla and families represented in the v4 16S SSU rDNA amplicons obtained from individual cod larvae, water, and live feed samples. At the phylum level, the microbiota is presented as bar graphs (bottom) and at the family level as a heat map (top) with the abundance of each family represented by a colored block as specified in the figure. Bars labeled D8, D17, D32, and D61 represent cod larva individuals at the ages 8, 17, 32, and 61 dph, respectively. Bars labeled W and F represent water and live feed samples, and the time of sampling is indicated in the label. Only taxa represented by a proportion of $\geq 1\%$ in at least one of the samples are shown. From Bakke et al. (2015)

were surprising, as diet has been identified to be a major determinant for fish intestinal microbiota. Moreover, numerically, the load of bacteria to the intestine is completely dominated by the microbiota from the feed (Vadstein et al. 2018a). A number of studies have indicated that diet influences the gut microbiota of fish [reviewed for marine fish in Egerton et al. (2018)]. Some studies on marine larvae based their conclusions on an observed change in the gut microbiota after onset of feeding (Jensen et al. 2004; Brunvold et al. 2007; Reid et al. 2009). However, the experimental design did not include controls, and it cannot be excluded that other factors contributed to these observations. It is interesting to note that a study on larvae of an unrelated marine fish species, orange-spotted grouper, came to conclusions similar to ours: the larval microbiota was not affected by change in live feed diet, and it was more similar to the water than the live feed microbiota (Sun et al. 2013).

3 Effect of Water Treatment and Water Microbiota on Microbiota and Gene Expression in Cod Larvae

High mortality of larvae and low reproducibility between replicate tanks are common in aquaculture hatcheries during the first weeks following hatching. These problems often are attributed to infections from opportunistic bacteria that proliferate under aquaculture conditions and cause disease in weakened hosts (Vadstein et al. 2004). In aquaculture, eggs are commonly surface disinfected, and the larvae depend on the microbial community in the surrounding water, and at mouth opening feed, for the primary colonization of the gut. The bacteria surrounding the fish are important as they contribute to normal development and protection against disease, by affecting the chemical water quality, but in certain unfortunate cases, they may lead to infectious disease and mortality.

The bacterial density and community composition in the rearing water of intensive aquaculture systems depend on water flow rates, bacteria in sources, and internal bacterial growth, loss, and competition. Bacterial densities can reach very high numbers in fish larval rearing units (>10^8 cells mL^{-1}) (Vadstein et al. 1993b). Different types of water treatment produce rearing water with significantly different species compositions (Attramadal et al. 2012a, b, 2014, 2016). The composition of the water microbial community seems to be more important for larval performance than absolute bacteria numbers, as long as numbers are not extreme (Munro et al. 1994; Salvesen et al. 1999; Verner-Jeffreys et al. 2004). The amount and types of bacteria that larvae are first exposed to may be especially important, as it has been shown to be harder to influence the microbial flora of the skin and digestive tract of individual larvae once colonized (Ringø and Vadstein 1998). A bad bacterial environment increases the likelihood for an unfortunate development for larvae during the first critical stage of microbial colonization.

Closed systems offer the possibility to influence the microbial environment of the larvae. In aquaculture, the focus has been on preventing specific pathogenic organisms from entering the fish farm by the use of hygiene barriers such as filtration and disinfection. However, a good microbial water quality is more than the absence of pathogenic microbes. The route to infection in larvae is shaped by the lack of a fully developed specific immune system, and specific pathogens are generally not considered as a major problem in hatcheries (Vadstein et al. 1993b; Vadstein 1997). On the contrary, outbreaks of disease often occur in apparent absence of known pathogens, and opportunistic microorganisms that become pathogenic when the host's resistance is lowered by environmental stress factors probably cause the majority of the diseases in marine fish larvae. A high share of opportunistic microbes is generally negative for the performance of the fish (Vadstein et al. 2004).

Disinfection of intake water is important to stop specific pathogens from entering the system. However, despite hygiene barriers, there will always be internal growth of bacteria in the system, both in the water and on the surfaces of tanks, pipes, and fish. The maximum number of bacteria that can be maintained in the system over time defines the microbial carrying capacity. The majority (more than 80%) of the bacteria in land-based aquaculture systems are heterotrophic (Michaud et al. 2006). The microbial carrying capacity is therefore primarily determined by the supply of available organic material. Thus, after disinfection, internal growth quickly restores the bacterial density to carrying capacity. The residence time of the water in the system and the growth rate of the bacteria determines whether the microbial density reaches the carrying capacity in the rearing tank.

According to the ecological theory of r-/K-selection (Mac Arthur and Wilson 1967), selective pressures drive succession in one of two generalized directions: Selection for opportunists (r-selection) occurs in unpredictable environments with empty niches, which favors the ability to reproduce quickly. Selection for specialists (K-selection), on the other hand, occurs in stable, predictable environments where the community is close to the carrying capacity of the system and where the ability to compete successfully for limited resources is favored. It has been hypothesized that larvae reared in water dominated by K-strategists, i.e., mature microbial communities, will perform better because they are less likely to encounter opportunistic microbes and develop detrimental host–microbe interactions (Vadstein et al. 1993b; Skjermo et al. 1997; Salvesen et al. 1999; Attramadal et al. 2014). To favor K-strategists, a stable supply of organic matter and a low bacteria loss rate by disinfection or water exchange are needed. A bacterial density close to the carrying capacity will ensure strong competition for substrates, and a long residence time for the water will allow growth of slow-growing specialist bacteria (Attramadal et al. 2014).

In an aquaculture system, daily operational routines like feeding and disinfection of the rearing water will create bacterial loss and excess resources. These conditions favor rapid-growing, opportunistic bacteria (Hess-Erga et al. 2010). Figure 7a shows how disinfection reduces the number of bacteria competing for the available resources and favors the subsequent proliferation of opportunists. Figure 7b shows how feeding leads to a rapid increase of available organic matter, resulting in advantageous conditions for the opportunistic bacteria.

Within hours following disinfection and feeding, uncontrolled regrowth leads to the bacterial density reaching the (new) carrying capacity of the rearing water. Stochastic events and selective forces shape the new community composition during the regrowth. This is the reason why two replicate fish tanks filled with disinfected water and fed a similar amount of feed reach a similar density of bacteria but often end up with significantly different community compositions.

Continuous and correct feeding combined with rapid and gentle removal of organic matter from the rearing water minimizes the carrying capacity and the density of bacteria and can to a certain degree counteract the destabilizing effects of feeding on the microbial community. In addition, it is possible to increase the stability and reproducibility of the microbial community between replicate rearing tanks by influencing the regrowth of bacteria following disinfection. This can be obtained by passing the water through a matured biofilter inhabited by a high density of bacteria that secures a strong competition for the nutrients and thus prevents opportunistic proliferation (Skjermo et al. 1997). This may be thought of as a microbial maturation of the water, leading to a K-selected microbial community

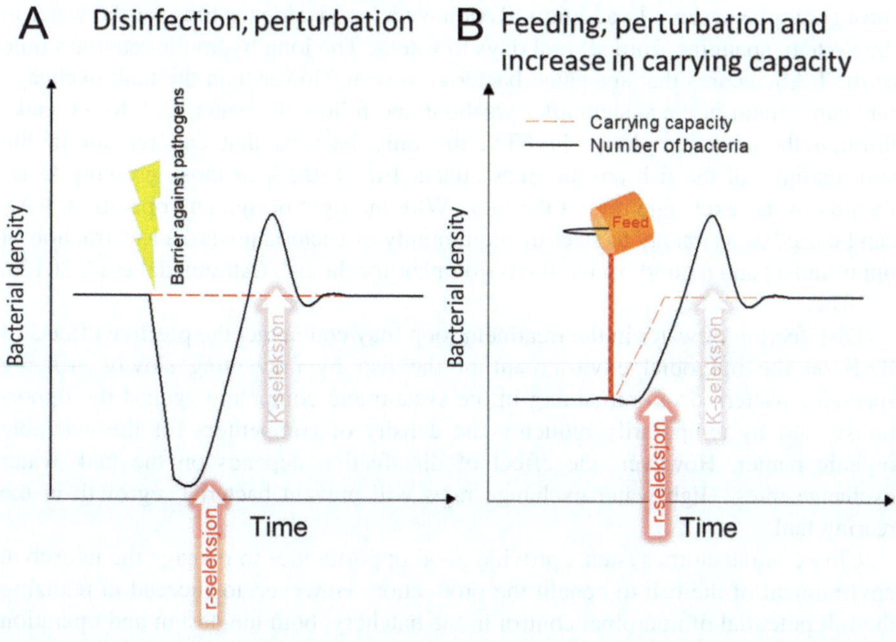

Fig. 7 Two examples of how current aquaculture practices promote r-selection increase the probability of opportunistic pathogens and consequently the probability of detrimental infections. (**a**) Disinfection of intake water is done as a barrier against introduction of pathogens in the system. However, disinfection also decreases bacterial numbers and makes dissolved organic matter more bioavailable that, under normal conditions, results in a bloom of fast-growing r-strategists. (**b**) Addition of feed to the system will increase the microbial carrying capacity in the rearing tanks directly and indirectly through defecation. This increase in carrying capacity will also result in r-selection. Notice that the water typically resides in the rearing tanks during the r-selection period and has already left the tanks when K-selection can take place. From De Schryver and Vadstein (2014)

and a reduced microbial carrying capacity in the rearing tanks, as the sessile biofilter bacteria consume significant amounts of the available substrate in the water on the way into the tank.

The development of the microbial community in the rearing tank depends on the incoming water and the residence time of the water in the tank. Low tank water exchange rates (hydraulic retention times of several hours) allow for significant internal growth of bacteria in the rearing tank. When the water exchange rate is low and the microbial carrying capacity of the rearing tank significantly exceeds that of the incoming water, uncontrolled regrowth of bacteria takes place in the surrounding of the fish larvae. To prevent this, the incoming water should be matured at the same carrying capacity as that in the rearing tank (Attramadal et al. 2016). This can be obtained by recirculating the rearing water. Recirculating aquaculture systems (RAS) contain biofilters (for the conversion of toxic ammonia to nitrate) with a large surface area for bacteria, and the microbial carrying capacity is relatively stable throughout the system. In addition, RAS have a long residence time for the water in the system, spanning from several days to weeks. The long hydraulic retention time of the RAS secures that specialist bacteria growing slower than the tank exchange rate can remain in the system after washout and follow the water back to the tanks through the treatment loop. In FTS, the only bacteria that can remain in the surroundings of the fish are the ones attached to surfaces or those growing faster than the water exchange rate of the tank. With the right design and operation, RAS can be used as a strategy to steer the community of bacteria toward a low fraction of opportunists and a good bacterial environment for the fish (Attramadal et al. 2012a, b, 2014).

Disinfection of water in the treatment loop may counteract the positive effects of RAS on the microbial environment of the fish by preventing slowly growing specialist bacteria from remaining in the system and competing against the opportunists and by temporarily reducing the density of competitors for the available organic matter. However, the effect of disinfection depends on the tank water exchange rates. High water exchange rates will prevent bacterial regrowth in the rearing tank.

Closed aquaculture systems provide good opportunities to manage the microbial environment of the fish to benefit the production. However, to succeed in realizing the full potential of microbial control in the hatchery, both the design and operation of the system must be considered in light of the system's microbial ecology. Water treatment, system design, and management can be used to control the selection pressure working on the microbial community in order to increase the fraction of harmless and beneficial bacteria at the expense of opportunists (Vadstein et al. 1993b). RAS with no disinfection has been suggested as a strategy for microbial control based on three features that promote K-selection: (1) the long retention time of water in the system, (2) the large surface area available for bacterial growth and competition in the biofilter, and (3) the stable microbial carrying capacity throughout the system (Attramadal et al. 2012b, 2014). In contrast, in FTS the change in microbial carrying capacity from clean intake water to significantly higher substrate levels in the rearing tanks creates r-selection and promotes opportunistic proliferation in tanks with low water exchange rates.

Here, we summarize the results from studies based on two separate first feeding experiments with cod larvae. We compared different system designs, RAS, FTS, and MMS (a flow-through system including a biofilter for microbial maturation; see below), and investigated the effects on the rearing water microbiota and on the larval survival, microbiota, and gene expression.

3.1 Summary of Research Results

We applied an experimental design similar to the one described above (see "Effect of Live Feed Diets on Cod Larval Microbiota"), but here we employed three distinct water treatment regimes (Attramadal et al. 2014): a recirculating aquaculture system

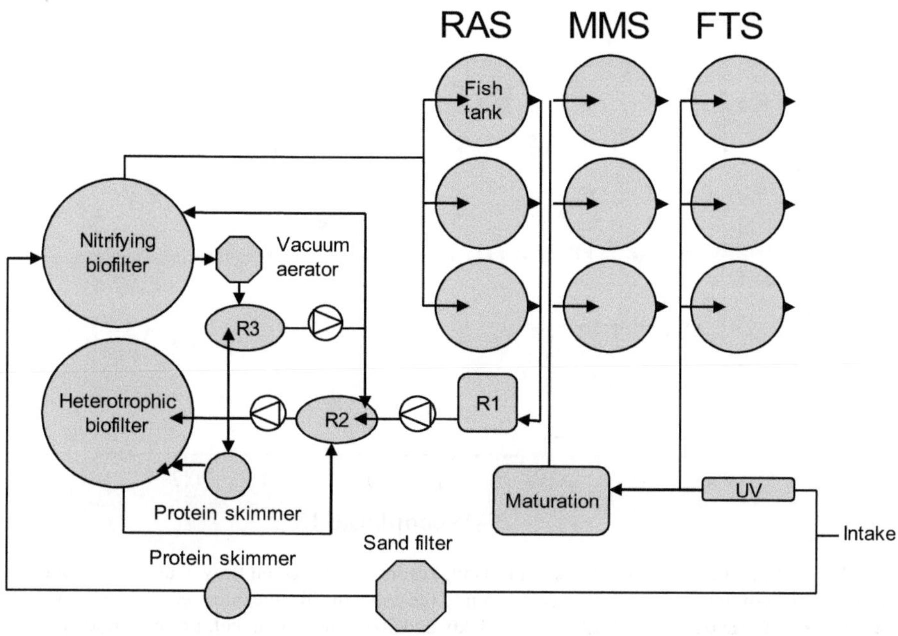

Fig. 8 Flow scheme of the water treatment in the RAS, the MMS, and the FTS. From Attramadal et al. (2014)

(RAS) with two biofilters, a flow-through system (FTS), and a flow-through system including a biofilter (to promote microbial maturation; MMS). All systems received the same UV-treated intake water. Figure 8 gives a schematic presentation showing the major steps in the water treatment. For each system, larvae were reared in three replicate rearing tanks of 160 liters at a density of 100 individuals per liter. From day 31 post hatching, all nine rearing tanks were operated as MMS, i.e., all rearing tanks received identical water. For details, see Attramadal et al. (2014).

In Attramadal et al. (2014), we analyzed only microbial communities associated with water samples (not cod larval samples), for the period when the RAS, FTS, and MMS rearing regimes were maintained. We found that the rearing water microbiota differed between the three systems, most profoundly between RAS and the other systems (Fig. 9). Moreover, we found that the rearing water microbiota was more stable over time in RAS than in FTS and MMS. The survival of larvae at 32 dph was higher in RAS and MMS than in the FTS system (Fig. 10). Thus, the RAS and MMS systems appeared to have positive effects on the larvae, and we suggested that this could be explained by a beneficial, K-selected microbiota in the rearing water.

Later, we also examined the larval microbiota from samples taken in the experiment: first by DGGE analysis and later Illumina sequencing of 16S SSU rRNA gene amplicons (results not previously published). The DGGE analyses showed that there were significant differences in larval microbiota between all three systems at 17 (Fig. 11a) and 30 dph and between RAS and the other two systems at 8 dph.

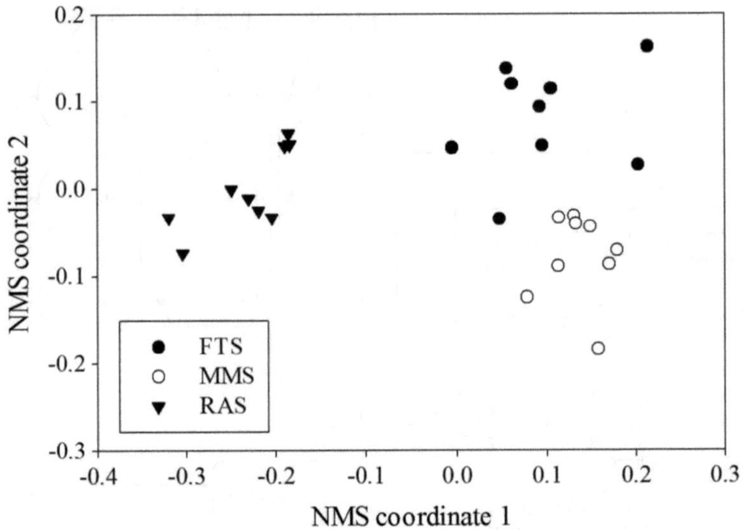

Fig. 9 NMS (nonmetric multidimensional scaling) ordination based on Bray–Curtis similarities for the microbial communities of the water going into the fish tanks for the three rearing systems (FTS, MMS, and RAS) at days 1, 4, 8, 12, 16, 19, 23, 26, and 30 of the experiment. From Attramadal et al. (2014)

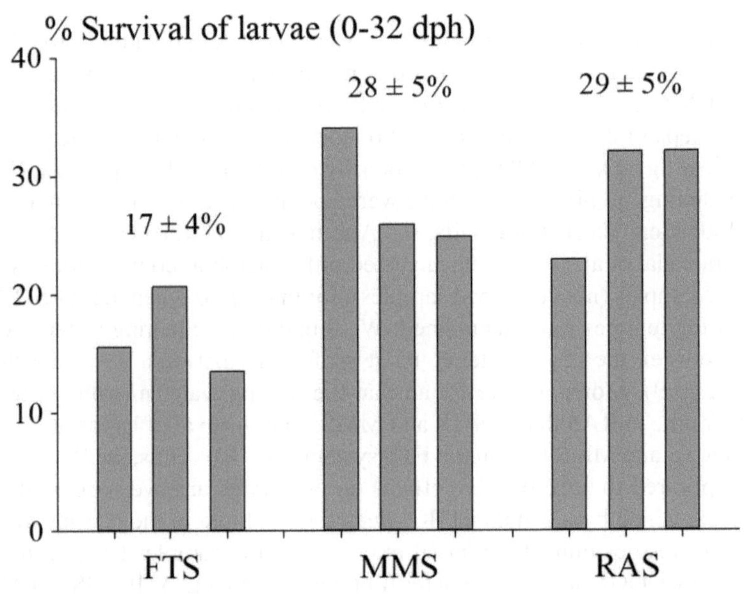

Fig. 10 Average survival ± SD of fish in RAS, FTS, and MMS during the live feed period (0–32 dph) in the three replicate tanks of each of the three systems. Each point represents on sample. From Attramadal et al. (2014)

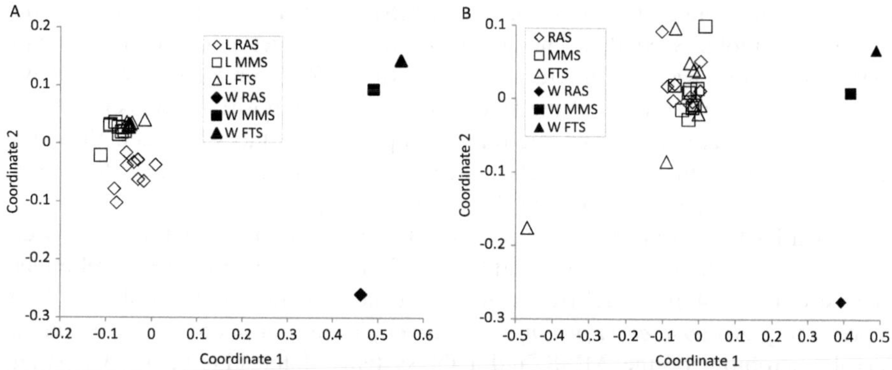

Fig. 11 NMS ordination based on Bray–Curtis similarities for the cod larval microbiota for the three rearing systems (FTS, MMS, and RAS) at 17 dph (**a**) and at 46 dph, when all rearing tanks had been operated as MMS and received identical rearing water for 16 days (**b**)

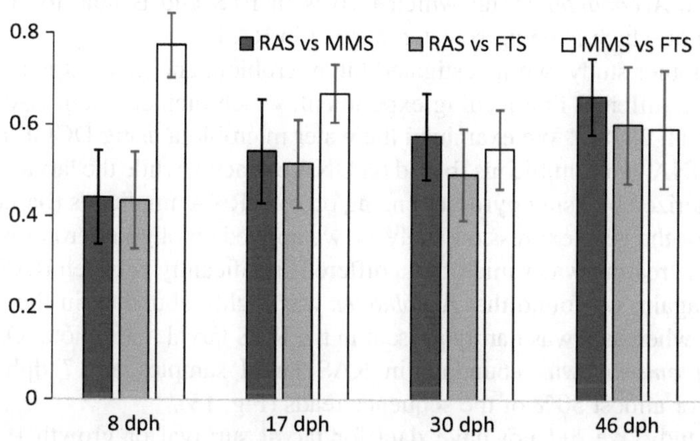

Fig. 12 Average Bray–Curtis similarities for comparisons of larval microbiota from different water treatment systems. Error bars indicate standard error (9 larvae per treatment and 81 Bray–Curtis similarities per comparison)

Average Bray–Curtis similarities for comparisons of cod larval microbiota between systems indicated that the FTS and MMS larval microbiota were more similar to each other than to the RAS larval microbiota at both 8 and 17 dph (Fig. 12), just as for the water microbiota (Attramadal et al. 2014).

At 46 dph, when all rearing tanks had received identical MMS water for 16 days, no differences in larval microbiota between any of the systems were detected (Fig. 11b). Illumina amplicon sequencing of the same samples confirmed these observations for the larval microbiota.

Thus, two different observations indicated that the larval microbiota is influenced by the rearing water microbiota: First, rearing cod larvae in water with distinct bacterial communities (RAS, FTS, and MMS) resulted in significantly different larval microbiota. Second, a change in the rearing water microbiota, supplying identical rearing water to all rearing tanks (from RAS or FTS to MMS), was followed by a convergence of the larval microbiota composition between systems. Hence, there seems to be an interrelationship between the water and the larval microbiota, even though the bacterial communities in the rearing water clearly differed from the larval microbiota (Fig. 11).

Which bacteria are responsible for the differences in larval microbiota between the systems, and how is the water and the larval microbiota interrelated? Preliminary analyses of the v4 16S SSU rRNA gene sequencing dataset indicate that an OTU representing *Arcobacter* (*Epsilonproteobacteria*) was far more abundant in the larval microbiota of the MMS and FTS systems. Interestingly, an *Arcobacter* OTU, with identical v4 16S rRNA gene sequence, was found to be highly abundant in the cod larval microbiota at 17 and 32 dph in the study where we examined the effect of different live feed diets on cod larval microbiota (see above). That first feeding experiment was performed in an FTS. This OTU might represent an opportunistic *Arcobacter* strain which thrives in FTS and is able to colonize the cod larvae but which is being selected against in RAS.

In yet another study, we investigated the microbiota and gene expression of cod larvae from a different first feeding experiment, which included both RAS and FTS (Vestrum et al. 2018b). We examined the water microbiota using DGGE analysis of 16S SSU rDNA gene amplicons (based on DNA extracts), while the larval microbiota was characterized by using pyrosequencing of 16S rRNA amplicons (based on RNA extracts). For the gene expression analysis, we applied an oligo microarray. Both the larval and the rearing water microbiota differed significantly between RAS and FTS. Strikingly, again, we found that *Arcobacter* was highly abundant in the FTS larval microbiota, whereas it was hardly present in the RAS larval microbiota. On the other hand, *Marinomonas* was abundant in RAS larval samples at 17 dph, where it accounted for almost 50% of the sequence reads (Fig. 13).

In this study, we did not have data for larval survival or growth [for reasons explained in Vestrum et al. (2018b)], but the gene expression analysis provided relevant information about larval performance. For larval samples from 8 and 13 dph, we did not find significant differences in gene expression between RAS and FTS. However, on 17 dph, 20 genes were differently expressed between the 2 systems. Of these, 19 were significantly upregulated in the larvae from the FTS, and 12 of these were found to be involved in processes coupled to pathogen recognition, infection, and immunity responses.

In summary, in these studies we demonstrated that aquaculture systems with different water treatment strategies selected for distinct rearing water microbiota. This affected the cod larval microbiota, survival, and gene responses. Our results are consistent with the hypothesis that RAS can be used to promote K-selection and microbial stability. This is obtained by maintaining a microbial load close to the carrying capacity of the system and by ensuring that the retention times for both bacteria and water in the system are long enough to prevent washout of beneficial bacteria. This appears to provide beneficial microbial environments for the larvae.

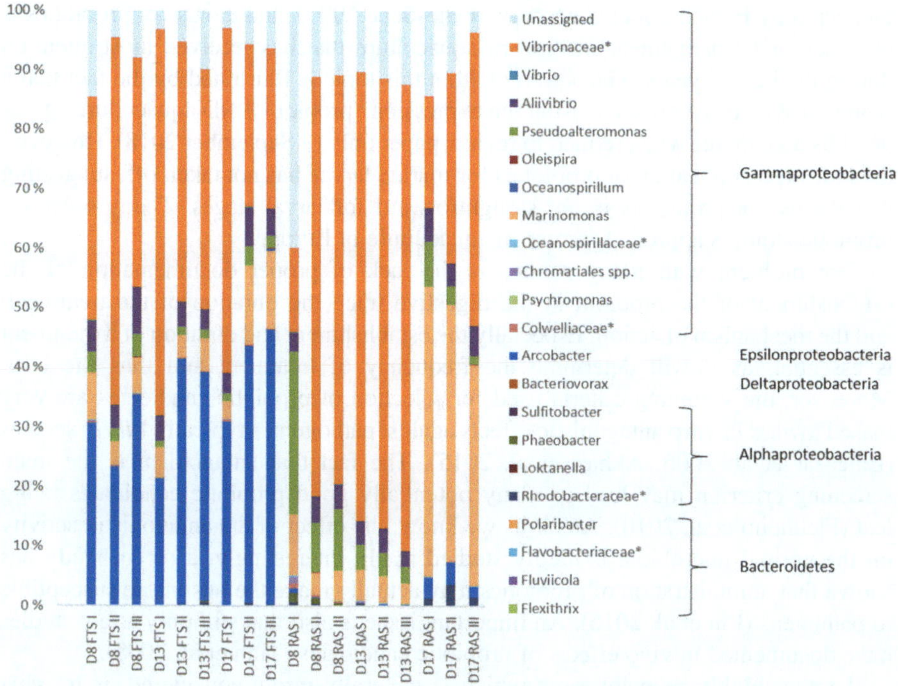

Fig. 13 Relative abundances of bacterial genera in cod larval samples. Taxa that could not be classified at the genus level are marked by asterisk. Each sample represents RNA extracts from 25 pooled cod larvae at the age 8 dph, 17 pooled cod larvae at 13 dph, and 10 pooled cod larvae at 17 dph (bars labeled D8, D13, and D17, respectively). Bars labeled FTS and RAS represent the flow-through system and recirculating aquaculture system, respectively. Only genera represented by a proportion of ≥1% in at least one of the samples are shown. From Vestrum et al. (2018b)

On the other hand, FTS may select for opportunistic bacteria, particularly if the retention time in the rearing tanks is relatively long and allows for blooming of rapid-growing strains, as for example, *Arcobacter*. Some of these may be able to colonize the fish and result in detrimental host–microbe interactions.

4 Potential for Manipulating the Cod Larval Microbiota by Probiotic Treatment

Probiotics are microorganisms that are administered to a host because of their beneficial effect. The concept, but not the name, was introduced by Nobel Prize winner Elie Metchnikoff in 1907, claiming that yoghurt consumption had a life extending effect due to the positive effects of lactic acid bacteria on the gut microbiota. The concept was applied in broiler production in the early 1970s (Nurmi and Rantala 1973) and came into more common use in the 1980s. In aquaculture, probiotics were pioneered by François-Joel Gatesoupe (Gatesoupe 1999), and the use of probiotics is the microbial management method in aquaculture that has received most attention during the last 25 years. The activity within this topic is illustrated by the fact that a search in Web of Knowledge using the expression "probiot* AND aquaculture" gives 967 hits and 85 hits when refined to review paper only (3 September 2018). However, the industrial application of probiotics for marine larvae has not taken off, suggesting that the use of probiotics is not straightforward for larval stages. Today only one probiotic strain is approved for use in aquaculture in Europe.

One problem with many studies is the lack of proper documentation of the establishment of the probiont in the digestive tract, the duration of the treatment, and the mechanism of action. Especially the establishment and duration of a treatment is essential, as it will determine the frequency of treatment and thus the cost. Moreover, the screening criteria used for selection of probiotic candidates are very biased toward in vitro antagonistic effects against pathogens – typically *Vibrio* species (Balcazar et al. 2006; Akhter et al. 2015). The fact that antagonism is the main screening criterion may lead to many potentially good probiotic candidates being lost (Fjellheim et al. 2010). As far as we know, the effect of the antagonistic activity on the normal microbiota is hardly studied at all. Interestingly, a recent study has shown that administration of probiotics may actually make the host more susceptible to pathogens (Liu et al. 2016). An important issue is also that still only a few studies have documented in vivo effects of probiotic candidates (Tinh et al. 2008).

Another highly relevant topic which is generally rarely considered, is to what extent a probiont is applicable practically and functionally throughout the whole larval stage. As described above, we found indications that the cod larval gut is a highly selective environment due to larvae–microbe and/or microbe–microbe interactions (Bakke et al. 2015). Further, we also found that there was a strong succession in the larval microbiota with time. This knowledge is relevant for the application of probiotics, as a possible consequence is that different probiotics have to be developed for different life stages. Often, probiotic candidates are tested only at one developmental stage, and thus this problem is generally not dealt with.

For larval stages, probionts can be administrated through two different routes: by addition to the water and by bioencapsulation in live feed. It is well documented that both rotifers and *Artemia* ingest bacteria (Vadstein et al. 1993a; Makridis and Vadstein 1999), and this method can be used to encapsulate bacteria in *Brachionus* and *Artemia* (Makridis et al. 2000a). Experiments have shown that both addition to water and bioencapsulation in live feed are efficient routes for transfer of probionts to larvae (Makridis et al. 2000b). Calculations suggest that numerically the most significant entry of microbes to the intestine of larvae is bacteria associated with the live feed (Vadstein et al. 2018a). This route constitutes more than 95% of the bacteria entering the digestive tract of Atlantic cod larvae, independent of rearing system. However, as shown above, the water microbiota has much higher similarity to the larval microbiota than the live feed microbiota. The mechanism for this is not known, but it might be related to qualitative differences in the entry route, higher alpha-diversity in water than in live feed, or that microbiota normally associated with the live feed has lower competitive ability in the digestive tract of the larvae. Still, the situation may be different for probionts, as several of the factors affecting the success in establishment in the digestive tract do not depend on the route of administration. A positive effect of probiotic bacteria in live feed may also be that they dilute or remove detrimental microbes from the live feed, similar to the effect shown to take place by addition of microalgae to *Artemia* cultures (Olsen et al. 2000).

Some of the considerations mentioned above, regarding the research on the use of probiotics in larval rearing, point to the lack of considering the host as an ecosystem with a rich microbiota of huge significance for the host (McFall-Ngai et al. 2013; Lerner et al. 2017). This is new knowledge, mostly established during the last two decades, and is due to a methodological revolution in molecular biology and imaging (Vestrum et al. 2018a). Even though most of this recent work on host-associated microbiomes is done on mammals, the broad functional repertoire and the microbial ecology perspective are also relevant for fish larvae (Vadstein et al. 2018a). The perturbation done by the addition of probiotics will definitely lead to a loss in diversity of the microbiota, and this may result in loss of stability and resilience and in increased risk of invasion by detrimental microbes (De Schryver and Vadstein 2014). We would argue that to secure maximum scientific output and impact, it is important that future research on probiotics implement a microbial ecology perspective.

4.1 Summary of Research Results

We designed an experiment to examine whether the microbiota of Atlantic cod larvae could be steered by introducing potential beneficial bacterial strains, which had previously been isolated from cod larvae, and further whether there are particular developmental stages where this can be obtained more easily (Skjermo et al. 2015). We used four bacterial strains, representing the genera *Microbacterium*, *Ruegeria*, *Pseudoalteromonas*, and *Vibrio*. The strains were added through live feed and rearing water during a treatment period of 24 hours. Treatments were performed for batches of larvae at different ages: 0, 2, 4, 8, 16, 30, and 45 dph. For each treatment group, we sampled rearing water and cod larvae at 0, 1, 2, 4, and 11 days after treatment, and the relative amounts of the added strains were determined by real-time PCR. Figure 14 summarizes the experimental design.

In general, the introduced strains were only detected in very low amounts in the larvae at the first day after treatment and then decreased to background levels. An exception was the *Microbacterium* strain, which was found to constitute as much as around 80% of the total 16S SSU rRNA gene copies [for a more detailed description of the quantification strategy, see Skjermo et al. (2015)] in the larval microbiota at the first day after treatment for the treatments performed at 2, 4, and 8 dph. However, also for this strain, the abundance rapidly decreased and was only found in very low levels in the larval samples 11 days after the treatment. Thus, even though the strains originated from cod larvae, they appeared to be only transiently present in the larval microbiota. Figure 15 summarizes the results for the *Microbacterium* and the *Pseudoalteromonas* strains in larval samples. There was, however, a tendency that

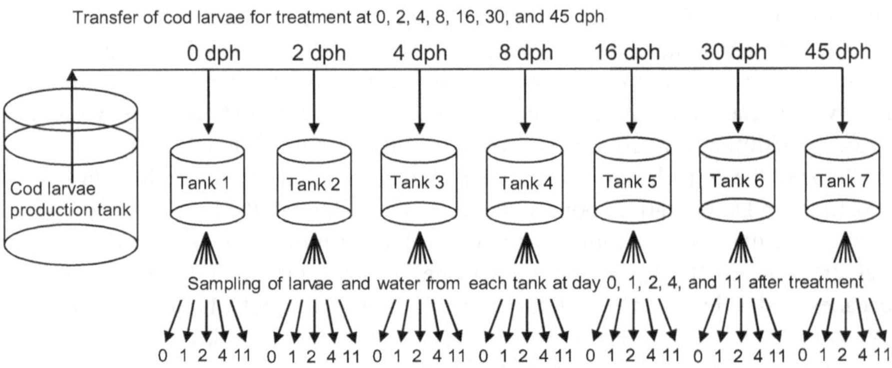

Fig. 14 Schematic presentation of experimental design for investigating the potential of introducing bacterial strains to the cod larval microbiota. For details, see Skjermo et al. (2015)

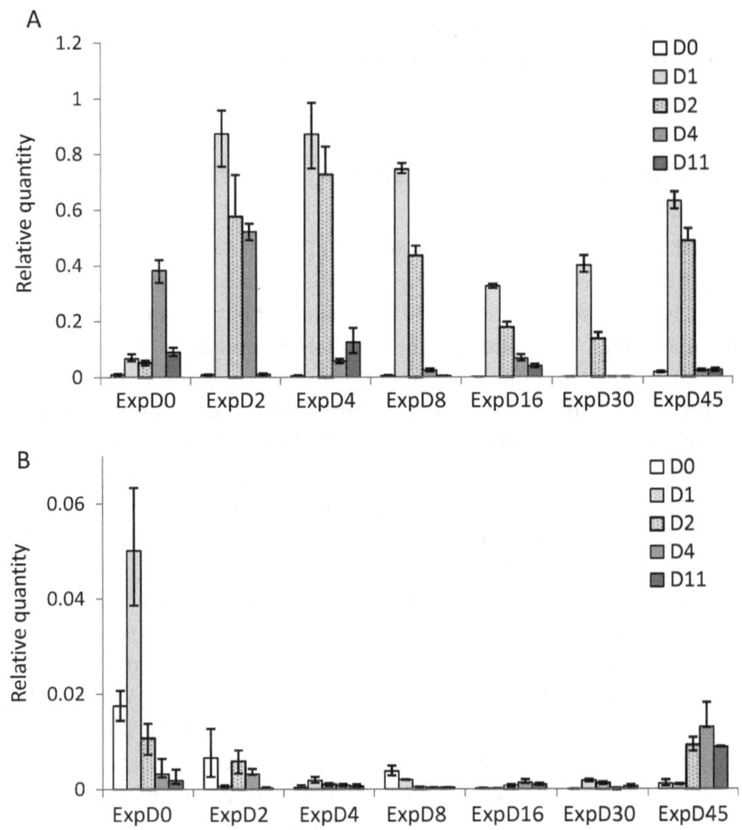

Fig. 15 Relative quantity of the bacterial strains *Microbacterium* (**a**) and *Pseudoalteromonas* (**b**) associated with cod larvae in different experimental tanks named ExpDn, where n denotes the day post hatching for treatment. Note that the y-axis differs on the plots. From Skjermo et al. (2015)

the strains persisted for a somewhat longer time in the larval microbiota for the last treatment at 45 dph. This may imply that the larvae are more susceptible to introduction of new strains in the microbiota at later developmental stages.

This study indicated that probiotic treatment aiming for the establishment of new strains in the cod larval microbiota is difficult, and it challenges the probiotic concept for the larval stage, unless the probiotics are continuously or repeatedly added to the fish larvae.

5 Summary and Conclusions

In previous research, we have shown that marine fish larvae are particularly vulnerable to poor microbial water quality. Not only specific pathogens, but also opportunistic bacteria, may cause mortality at early life stages. The studies on the cod larval microbiota reviewed here, identified the water microbiota and selection in the host as two important factors shaping the larval microbiota. We applied similar experimental designs and sampling times to investigate effects of diet and water treatment. When comparing the effects on the larval microbiota at 17 dph, the larval microbiota clearly differed between water treatment systems, but not between different feeding regimes (Fig. 16). This was surprising, because diet has previously been identified as an important factor in shaping the gut microbiota in fish. Characterization of the microbiota in developing cod larvae revealed major changes over time, which could not be explained by changes in environmental factors. This suggests that factors in the host, probably associated with developmental changes in the intestine and the immune system, are important in structuring the larval microbiota. We further found that it was difficult to manipulate the cod larval microbiota by introducing probiotic bacteria through water and live feed. This finding can be explained by a strong selection in the host (host–microbe and/or microbe–microbe interactions). An obvious question is whether it is possible to steer the cod larval microbiota toward a healthy, beneficial state. Based on our current knowledge, we suggest that the best way to manage the microbiota of the fish larvae is through managing the rearing water microbiota. While microbial management in aquaculture systems traditionally has been focusing on achieving low bacterial numbers and keeping specific pathogens out of the system, we strongly advocate a focus on the microbial community composition of the water. At this stage, we are not able to identify which microbes are antagonistic, neutral, or beneficial for the fish larvae. However, by applying principles from theoretical and microbial ecology, we have shown that it is possible to predict conditions that create healthy microbial environments. Stable, K-selected environments promote proliferation of beneficial bacteria in the rearing water. Moreover, according to ecological theory, under such conditions opportunistic pathogens will not thrive.

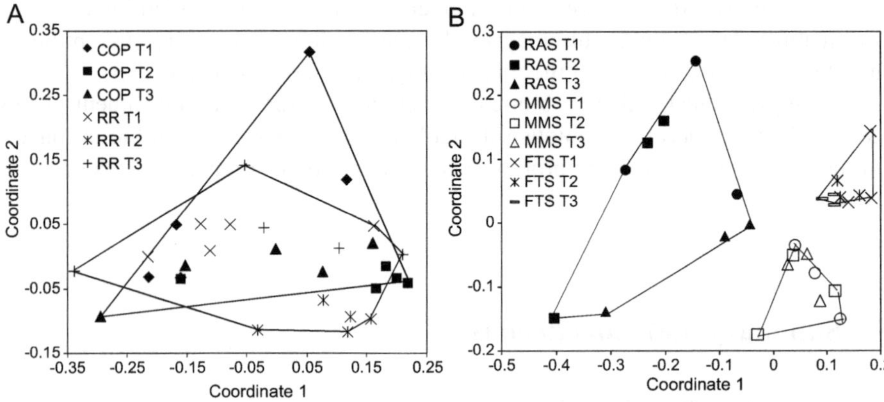

Fig. 16 Nonmetric MDS plots based on Bray–Curtis similarities for cod larval microbiota at 17 dph. (**a**) Microbiota from 30 larvae representing the diets COP (copepods fed algae) and RR (rotifers fed algae); 5 individuals from each replicate rearing tank (based on data from Bakke et al. 2013). (**b**) Microbiota from 27 larvae representing the rearing water systems FTS, MMS, and RAS; 3 individuals from each replicate tank. From Vadstein et al. (2018a)

References

Akhter N, Wu B, Memon AM, Mohsin M (2015) Probiotics and prebiotics associated with aquaculture: a review. Fish Shellfish Immunol 45:733–741

Attramadal KJK, Minniti G, Oie G, Kjorsvik E, Ostensen MA, Bakke I, Vadstein O (2016) Microbial maturation of intake water at different carrying capacities affects microbial control in rearing tanks for marine fish larvae. Aquaculture 457:68–72

Attramadal KJK, Oie G, Storseth TR, Alver MO, Vadstein O, Olsen Y (2012a) The effects of moderate ozonation or high intensity UV-irradiation on the microbial environment in RAS for marine larvae. Aquaculture 330:121–129

Attramadal KJK, Salvesen I, Xue RY, Oie G, Storseth TR, Vadstein O, Olsen Y (2012b) Recirculation as a possible microbial control strategy in the production of marine larvae. Aquac Eng 46:27–39

Attramadal KJK, Truong TMH, Bakke I, Skjermo J, Olsen Y, Vadstein O (2014) RAS and microbial maturation as tools for K-selection of microbial communities improve survival in cod larvae. Aquaculture 432:483–490

Bakke I, Coward E, Andersen T, Vadstein O (2015) Selection in the host structures the microbiota associated with developing cod larvae (*Gadus morhua*). Environ Microbiol 17:3914–3924

Bakke I, Skjermo J, Vo TA, Vadstein O (2013) Live feed is not a major determinant of the microbiota associated with cod larvae (*Gadus morhua*). Environ Microbiol Rep 5:537–548

Balcazar JL, de Blas I, Ruiz-Zarzuela I, Cunningham D, Vendrell D, Muzquiz JL (2006) The role of probiotics in aquaculture. Vet Microbiol 114:173–186

Brunvold L, Sandaa RA, Mikkelsen H, Welde E, Bleie H, Bergh O (2007) Characterisation of bacterial communities associated with early stages of intensively reared cod (*Gadus morhua*) using denaturing gradient gel electrophoresis (DGGE). Aquaculture 272:319–327

Datta MS, Almada AA, Baumgartner MF, Mincer TJ, Tarrant AM, Poiz MF (2018) Inter-individual variability in copepod microbiomes reveals bacterial networks linked to host physiology. ISME J 12:2103–2113

De Schryver P, Vadstein O (2014) Ecological theory as a foundation to control pathogenic invasion in aquaculture. ISME J 8:2360–2368

Egerton S, Culloty S, Whooley J, Stanton C, Ross RP (2018) The gut microbiota of marine fish. Front Microbiol 9:873

Fjellheim AJ, Klinkenberg G, Skjermo J, Aasen IM, Vadstein O (2010) Selection of candidate probionts by two different screening strategies from Atlantic cod (*Gadus morhua* L.) larvae. Vet Microbiol 144:153–159

Gatesoupe FJ (1999) The use of probiotics in aquaculture. Aquaculture 180:147–165

Hess-Erga O-K, Blomvagnes-Bakke B, Vadstein O (2010) Recolonization by heterotrophic bacteria after UV irradiation or ozonation of seawater; a simulation of ballast water treatment. Water Res 44:5439–5449

Jensen S, Ovreas L, Bergh O, Torsvik V (2004) Phylogenetic analysis of bacterial communities associated with larvae of the Atlantic halibut propose succession from a uniform normal flora. Syst Appl Microbiol 27:728–736

Kanther M, Rawls JF (2010) Host-microbe interactions in the developing zebrafish. Curr Opin Immunol 22:10–19

Kelly C, Salinas I (2017) Under pressure: interactions between commensal microbiota and the Teleost immune system. Front Immunol 8:559

Lerner A, Neidhofer S, Matthias T (2017) The gut microbiome feelings of the brain: a perspective for non-microbiologists. Microorganisms 5

Liu Z, Liu WS, Ran C, Hu J, Zhou ZG (2016) Abrupt suspension of probiotics administration may increase host pathogen susceptibility by inducing gut dysbiosis. Sci Rep 6:23214

Llewellyn MS, Boutin S, Hoseinifar SH, Derome N (2014) Teleost microbiomes: the state of the art in their characterization, manipulation and importance in aquaculture and fisheries. Front Microbiol 5:207

Mac Arthur RH, Wilson EO (1967) The theory of island biogeography, Monographs in population biology. Princeton University Press, Princeton

Makridis P, Fjellheim AJ, Skjermo J, Vadstein O (2000a) Control of the bacterial flora of Brachionus plicatilis and *Artemia franciscana* by incubation in bacterial suspensions. Aquaculture 185:207–218

Makridis P, Jon Fjellheim A, Skjermo J, Vadstein O (2000b) Colonization of the gut in first feeding turbot by bacterial strains added to the water or bioencapsulated in rotifers. Aquac Int 8:367–380

Makridis P, Vadstein O (1999) Food size selectivity of *Artemia franciscana* at three developmental stages. J Plankton Res 21:2191–2201

McFall-Ngai M, Hadfield MG, Bosch TCG, Carey HV, Domazet-Loso T, Douglas AE, Dubilier N, Eberl G, Fukami T, Gilbert SF, Hentschel U, King N, Kjelleberg S, Knoll AH, Kremer N, Mazmanian SK, Metcalf JL, Nealson K, Pierce NE, Rawls JF, Reid A, Ruby EG, Rumpho M, Sanders JG, Tautz D, Wernegreen JJ (2013) Animals in a bacterial world, a new imperative for the life sciences. Proc Natl Acad Sci U S A 110:3229–3236

Michaud L, Blancheton JP, Bruni V, Piedrahita R (2006) Effect of particulate organic carbon on heterotrophic bacterial populations and nitrification efficiency in biological filters. Aquac Eng 34:224–233

Munro PD, Barbour A, Birkbeck TH (1994) Comparison of the gut bacterial flora of start-feeding larval turbot reared under different conditions. J Appl Bacteriol 77:560–566

Nemergut DR, Schmidt SK, Fukami T, O'Neill SP, Bilinski TM, Stanish LF, Knelman JE, Darcy JL, Lynch RC, Wickey P, Ferrenberg S (2013) Patterns and processes of microbial community assembly. Microbiol Mol Biol Rev 77:342–356

Nurmi E, Rantala M (1973) New aspects of salmonella infection in broiler production. Nature 241:210–211

Olsen AI, Olsen Y, Attramadal Y, Christie K, Birkbeck TH, Skjermo J, Vadstein O (2000) Effects of short term feeding of microalgae on the bacterial flora associated with juvenile *Artemia franciscana*. Aquaculture 190:11–25

Reid HI, Treasurer JW, Adam B, Birkbeck TH (2009) Analysis of bacterial populations in the gut of developing cod larvae and identification of Vibrio logei, Vibrio anguillarum and Vibrio splendidus as pathogens of cod larvae. Aquaculture 288:36–43

Reitan KI, Natvik CM, Vadstein O (1998) Drinking rate, uptake of bacteria and microalgae in turbot larvae. J Fish Biol 53:1145–1154

Ringø E, Vadstein O (1998) Colonization of Vibrio pelagius and Aeromonas caviae in early developing turbot (*Scophthalmus maximus* L.) larvae. J Appl Microbiol 84:227–233

Rosenberg E, Zilber-Rosenberg I (2011) Symbiosis and development: the hologenome concept. Birth Defects Res C Embryo Today 93:56–66

Salvesen I (1999) Microbial ecology in early life stages of marine fish: development and evaluation of methods for microbial management in intensive larviculture. Norwegian University of Science and Technology, Trondheim

Salvesen I, Skjermo J, Vadstein O (1999) Growth of turbot (*Scophthalmus maximus* L.) during first feeding in relation to the proportion of r/K-strategists in the bacterial community of the rearing water. Aquaculture 175:337–350

Skjermo J, Bakke I, Dahle SW, Vadstein O (2015) Probiotic strains introduced through live feed and rearing water have low colonizing success in developing Atlantic cod larvae. Aquaculture 438:17–23

Skjermo J, Salvesen I, Oie G, Olsen Y, Vadstein O (1997) Microbially matured water: a technique for selection of a non-opportunistic bacterial flora in water that may improve performance of marine larvae. Aquac Int 5:13–28

Skjermo J, Vadstein O (1993) Characterization of the bacterial-flora of mass cultivated brachionus-plicatilis. Hydrobiologia 255:185–191

Sun YX, Yang HL, Ling ZC, Ye JD (2013) Microbial communities associated with early stages of intensively reared orange-spotted grouper (*Epinephelus coioides*). Aquac Res. https://doi.org/10.1111/are.12167

Tinh NTN, Dierckens K, Sorgeloos P, Bossier P (2008) A review of the functionality of probiotics in the larviculture food chain. Mar Biotechnol 10:1–12

Vadstein O (1997) The use of immunostimulation in marine larviculture: possibilities and challenges. Aquaculture 155:401–417

Vadstein O, Attramadal KJK, Bakke I, Forberg T, Olsen Y, Verdegem M, Giatsis C, Skjermo J, Aasen IM, Gatesoupe F-J, Dierckens K, Sorgeloos P, Bossier P (2018a) Managing the microbial community of marine fish larvae: a holistic perspective for larviculture. Front Microbiol 9:1820

Vadstein O, Attrtamadal KJK, Bakke I, Olsen Y (2018b) K-selection as microbial community management strategy: a method for improved viability of larvae in aquaculture. Front Microbiol 9:2730

Vadstein O, Bergh O, Gatesoupe FJ, Galindo-Villegas J, Mulero V, Picchietti S, Scapigliati G, Makridis P, Olsen Y, Dierckens K, Defoirdt T, Boon N, De Schryver P, Bossier P (2013) Microbiology and immunology of fish larvae. Rev Aquac 5:S1–S25

Vadstein O, Mo TA, Bergh Ø (2004) Microbial interactions, prophylaxis and diseases. In: Moksness E, Kjørsvik E, Olsen Y (eds) Culture of old-water fishes. Blackwell, Bath, pp 28–72

Vadstein O, Oie G, Olsen Y (1993a) Particle-size dependent feeding by the rotifer brachionus-plicatilis. Hydrobiologia 255:261–267

Vadstein O, Oie G, Olsen Y, Salvesen I, Skjermo J, Skjakbraek G (1993b) A strategy to obtain microbial control during larval development of marine fish. Fish Farm Technol:69–75

Vandepoele K, De Vos W, Taylor JS, Meyer A, Van de Peer Y (2004) Major events in the genome evolution of vertebrates: Paranome age and size differ considerably between ray-finned fishes and land vertebrates. Proc Natl Acad Sci U S A 101:1638–1643

Vellend M (2010) Conceptual synthesis in community ecology. Q Rev Biol 85:183–206

Verner-Jeffreys DW, Shields RJ, Bricknell IR, Birkbeck TH (2004) Effects of different water treatment methods and antibiotic addition on larval survival and gut microflora development in Atlantic halibut (*Hippoglossus hippoglossus* L.) yolk-sac larvae. Aquaculture 232:129–143

Vestrum RI, Luef B, Forberg T, Bakke I, Vadstein O (2018a) Investigating fish larvae-microbe interactions in the 21st century: old questions studied with new tools. In: Yúfera M (ed) Emerging issues in fish larvae research. Springer, Cham

Vestrum RI, Attramadal KJK, Winge P, Li K, Olsen Y, Bones AM, Vadstein O, Bakke I (2018b) Rearing water treatment induces microbial selection influencing the microbiota and pathogen associated transcripts of cod (*Gadus morhua*) larvae. Front Microbiol 9:851

Wang AR, Ran C, Ringø E, Zhou ZG (2017a) Progress in fish gastrointestinal microbiota research. Rev Aquac:1–15

Wang B, Yao M, Lv L, Ling Z, Li L (2017b) The human microbiota in health and disease. Engineering 3:71–82

Insights into Mussel Microbiome

J. A. Rubiolo, L. M. Botana, and P. Martínez

Abstract Mussels along with clams and oyster represent the most important worldwide mollusk production. Among mussels, *Mytilus chilensis* and *Mytilus galloprovincialis* are the most important species, exceeding 300,000 tons, but other species like *Mytilus edulis* and *Mytilus californianus* have significant production. In this chapter, we focus on the relevance of mussel microbiota for aquaculture and, specifically, for human health. In some sections, due to the limited documentation available, information has been expanded to studies in other mollusks, such as oysters and clams, as a significant reference for future studies on mussels. Anyway, information on mollusk microbiota lags behind other aquaculture species, particularly fish, and this limitation is discussed in this chapter. Then, the demand for further work is highlighted considering the benefits of microbiota control for mollusk production and health.

1 Mussel vs. Environment Microbiota: Relevance for Production and Human Health

Mussels comprise a diverse family (Mytilidae) of freshwater and marine bivalves. These mollusks are suspension feeders capable of filtering large volumes of water through their gills to fulfill their food requirements. The organic particles that can be used as food include microalgae (e.g., phytoplankton; Wright et al. 1982), bacteria (Langdon and Newell 1990), vascular plants, or macroalgae detritus (Kreeger et al. 1988), and organic aggregates (Alber and Valiela 1994). Quite a few marine species

J. A. Rubiolo (✉) · P. Martínez
Departamento de Zoología, Genética y Antropología Física, Facultad de Veterinaria, Universidad de Santiago de Compostela, Lugo, Spain
e-mail: ja.rubiolo@usc.es

L. M. Botana
Departamento de Farmacología, Facultad de Veterinaria, Universidad de Santiago de Compostela, Lugo, Spain

N. Derome (ed.), *Microbial Communities in Aquaculture Ecosystems*,
https://doi.org/10.1007/978-3-030-16190-3_5

of the genera *Mytilus* (blue mussels) and *Perna* (greenshell) pertaining to the highly diverse family Mytilidae have been investigated to date, despite their important ecological and/or commercial value. Mussel growth rate and reproduction are affected by the amount and quality of food (Hawkins and Bayne 1992) with a direct impact on its production performance and commercial value. Moreover, the relationship of bacteria and benthic filter feeders may be functionally important to aquatic ecosystems (Prins et al. 1998). The filtering activity of these bivalves may affect the levels of the bacterioplankton, including heterotrophic bacteria (Cole et al. 1988), and, thus, may influence microbial community in terms of both abundance and biodiversity (Kiorboe et al. 1981).

The microbiota of bivalves (mostly based on *Crassostrea gigas*) was originally examined by classic microbiological methods in the search for pathogenic microorganisms that could cause diseases associated with human consumption (Kaysner et al. 1989; Kaspar and Tamplin 1993; Høi et al. 1998). Nonindigenous pathogenic bacteria that can be encountered in mussels include *Escherichia coli*, *Vibrio* spp., *Salmonella* spp., and *Shigella* spp. (Iwamoto et al. 2010). These bacteria eventually concentrate in bivalves during the process of filter feeding when they live in waters naturally contaminated with these pathogens or contaminated with sewage-polluted waters. Nevertheless, studies on bacterial seasonal composition, integrating the natural microbiota of bivalve mollusks and that of the surrounding environment, are scarce and with very diverse outcomes. It has been suggested that bivalve microbiota is very similar to that of the surrounding waters (Sugita et al. 1981) and, therefore, that mussel microbiota would reflect the increasing concentration of Enterobacteriaceae and Aeromonadaceae related to anthropogenic impact (Beleneva et al. 2003). Notwithstanding, it has been shown that the most frequent bacterial genera in *C. gigas* living in polluted areas are *Pseudomonas*, *Vibrio*, *Acinetobacter*, and *Aeromonas*, *Vibrio* being even more abundant than in surrounding waters (Kueh and Chan 1985; Prieur et al. 1990). A similar observation has more recently been reported in mussels, whose microbiota reflected that of the surrounding waters but with higher concentration of heterotrophic bacteria (Cavallo et al. 2009). Another study comparing the oyster and mussel microbiomes using 16S rDNA pyrosequencing showed that the microbial communities associated with both species were significantly different from the surrounding environment. Despite more data being needed, it appears that bivalve microbiome largely depends on the waters where they live, but with important nuances depending on the mollusk species which determines a selective enrichment of some bacterial genera, such as *Vibrio* and *Pseudomonas* (Vezzulli et al. 2018). Information at hand suggests that despite the environmental background greatly influencing mollusk microbiome taxonomic composition, the process of selective filtering determines an increasing concentration of specific bacteria. Accordingly, if mollusks develop in polluted waters with human pathogenic bacteria, these will be to some degree accumulated and, if not appropriately managed, will end up as part of the human diet. Furthermore, the bacterial strains present in the water where mussel lives might be pathogenic for mollusk themselves decreasing their production performances.

2 Mussel Microbiota Diversity

The microbiota of several mussel tissues have been characterized (Fig. 1) in bivalves collected in different locations. Due to its influence, the growth in a particular environment is expected to shape at least partially the mussel-associated microbiome to different tissues, modifying the taxonomic diversity and relative abundances of bacterial strains, despite the presence of tissue-specific core bacterial species. In the next subsections, the microbiota of hemolymph, gut, shell, and gill are described based on available studies.

2.1 Hemolymph Microbiota

While the circulatory system of healthy vertebrates is sterile, it has been shown that the hemolymph of many healthy aquatic invertebrates contains bacteria. The immunity provided by hemolymph involves cell-mediated and humoral systems that coordinately protect the organism from invading pathogens. The cell-mediated system depends on hemocytes capable of killing microbes through phagocytosis and/or cytotoxic reactions (release of lysosomal enzymes, antimicrobial peptides, and reactive oxygen species—ROS) (Cheng 1975; Hine 1999; Canesi et al. 2002; Pila et al. 2016). The noncellular portion of hemolymph contains specific proteins such as lectins, acid phosphatase, lysozyme, and antimicrobial peptides (Leippe and Renwrantz 1988; Pipe 1990; Leclerc 1996; Carballal et al. 1997; Mitta et al. 2000; Muroga and Takahashi 2007).

Due to the highly variable conditions observed in water environments, seawater shows a huge diversity of bacteria composition. Microorganisms can enter into the mollusk hemolymph since bivalves have an open circulatory system where the

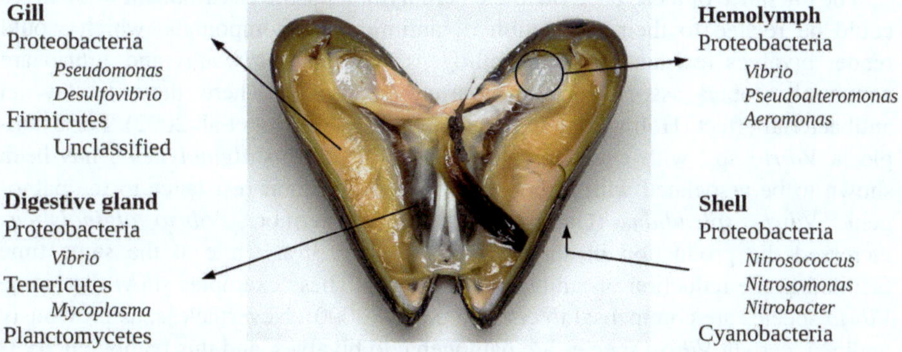

Fig. 1 Highly represented bacterial phyla (when known most abundant genus is indicated) in mussel tissues. In the case of hemolymph, the extraction site is indicated by the arrow

hemolymph leaves the arteries for bathing all organs and then returns to the heart through sinuses and gills. Bacterial entrance to the hemolymph occurs through invasive interactions or after ingestion during normal filter feeding. In principle, the hemolymph blood cells, the hemocytes, and other soluble factors coordinately provide protection from invading microorganisms (Renwrantz 1983; Matranga 1996). Using traditional microbiological culture techniques, it was shown that healthy horse mussel (*Modiolus modiolus*), swan mussel (*Anodonta cygnea*), and Pacific oyster (*Crassostrea gigas*) hemolymph are colonized by bacteria with concentrations ranging from 1.10^2/mL to 6×10^2/mL approximately. The dominant genera observed were *Vibrio*, *Alteromonas*, *Pseudomonas*, and *Aeromonas* in the case of oysters and horse mussel and *Vibrio* and *Aeromonas* in swan mussel (Olafsen et al. 1993; Antunes et al. 2010). These analyses point to a certain selection of bacterial uptake by bivalves with preference for the *Vibrio* genus. Some species of this genus are responsible for different diseases observed in bivalves (Elston et al. 1981; Jeffries 1982; Paillard et al. 2004; Romalde et al. 2014; Romero et al. 2014; Travers et al. 2015; Dubert et al. 2017), which suggests that bivalves can cope to a certain extent with these pathogens that could eventually become harmful in the presence of stressful conditions, turning commensal species into pathogenic (Grimes et al. 1984). It is worth mentioning that *Vibrio* accumulation transforms bivalves into vectors that can spread human pathogenic Vibrios (*V. cholerae*, *V. vulnificus*, *V. parahaemolyticus*) (Hervio-Heath et al. 2002; Collin and Rehnstam-Holm 2011; Raszl et al. 2016). In a recent study, the microbiomes of *C. gigas* and *M. galloprovincialis* co-cultivated in the same farm were compared using 16S rDNA pyrosequencing (Vezzulli et al. 2018). Since both species were under the same environmental conditions, the differences or similarities observed would be expected to be species-specific. However, the hemolymph microbiota was highly similar in both bivalves and dominated by *Vibrio* and *Pseudoalteromonas*. This observation gives further support to these bacterial genera as dominant members of oyster and mussel hemolymph, being able to persist and survive in this antibacterial environment.

The tolerance of these bacteria to the hemolymph hostile environment of bivalves could be related to their production of antimicrobial compounds, which would render bivalves resistant to other pathogens. *Pseudoalteromonas* and *Vibrio* are commonly found associated with marine eukaryotes where they display an antibacterial effect (Holmström and Kjelleberg 1999; Engel et al. 2002). For example, a *Vibrio* sp., with characteristics resembling *Vibrio alginolyticus*, has been shown to be associated with turbot larvae increasing their resistance to the pathogenic *Vibrio splendidus* (Gatesoupe 1997). Furthermore, *Vibrio alginolyticus* increased the production in shrimp hatcheries by 35% while at the same time facilitating the reduction of antibiotic treatments. These examples show that some *Vibrio* species are commensal (Verschuere et al. 2000). Nevertheless, as previously outlined, certain *Vibrio* species are pathogenic to bivalves, and this pathogenicity is higher in oysters than in mussels, which are more robust to microbial infections (Labreuche et al. 2006). Accordingly, it was found that oysters contain a higher proportion of *Vibrio* in their hemolymph microbiota as compared to mussels, which implies a higher risk in stressful conditions where certain *Vibrio* species can become pathogenic for the host. Furthermore, *V. aestuarianus*, frequently detected in oysters, was not observed in mussels. The presence of the soluble components (MgEP) in the Mediterranean mussel (*Mytilus galloprovincialis*) hemolymph, which mediates adhesion, internalization, and killing of *V. aestuarianus* by hemocytes (Balbi et al. 2013; Pezzati et al. 2015; Canesi et al. 2016), could be responsible for this difference. Overall, it appears that bivalve diseases, always depending on the species, may be due to increasing pathogenic bacteria in the surrounding water as a consequence of changes in environmental factors or to particular conditions in the hosts which could turn commensal bacteria into pathogenic.

2.2 Gut Microbiota

The few studies carried out to assess bacterial diversity of the mollusk digestive gland by marker sequencing have shown the notable differentiation of the hepatopancreas microbiome regarding other organs. So Vezzulli et al. (2018) showed that Proteobacteria, mostly *Desulfovibrio and Vibrio* genera, represented the major fraction of the gut microbiota (~72%) of mussels from Ligurian Sea (NW Italy), which in addition showed a minor representation of *Pseudoalteromonas* genus (~3%). This means that the phylum Proteobacteria might represent two-thirds of the microbiota of mussel hepatopancreas. In another study, Rubiolo et al. (2018) showed that the hepatopancreas bacteriome of Mediterranean mussel used for human consumption from three different Galician locations (NW Spain) was mainly composed of Proteobacteria (range: ~18–25%), Tenericutes (~4–20%), Planctomycetes (~10–22%), and Cyanobacteria (~10–43%), phyla that had been previously reported in this species by Craft et al. (2010) and in *M. califonianus* by Pfister et al. (2010). However, an important variation in the representation of these phyla was detected depending on the mussel collection location, even after depuration, pointing toward the ability of environmental conditions to reshape the microbiota taxonomic composition. As in previous studies Proteobacteria were importantly represented in this case by Alpha- and Gamma-proteobacteria, the latter including several nonindigenous human pathogenic species like *Pseudomonas* spp., *E. coli*, *Vibrio* spp., and *Salmonella* spp. usually found in mussels living in sewage-contaminated waters. Nevertheless, the Alpha-proteobacteria class, involved in photosynthesis, nitrogen fixation, ammonia oxidation, and methylotrophy (Viollier and Shapiro 2004), proportionally increased when mussels were treated with clean water for filtering (see below). Metagenomic analyses highlighted the importance of the environment to explain not only the diversity of bacteria in mussel gut but also the bacterial metabolism itself. Depending on the harvesting location, there were important differences in functional repertories involved in the xenobiotic metabolism and the synthesis of secondary metabolites. These results show how bacteria respond to different environmental pressures, a fact that could also help mussels to cope with them, although additional work will be necessary for its experimental demonstration.

In the aforementioned studies, the effect of depuration on mussel gut microbiota was also determined. This process, under conditions that maximize shellfish filtering activity, is an efficient method to eliminate potentially harmful microorganisms from bivalves (Richards 1991) and was originally developed to deal with a number of shellfish-associated typhoid disease outbreaks (Roderick and Schneider 1994). This process, performed after collection and before commercialization of farmed bivalves, involves placing mussels in tanks where they resume filter feeding in the presence of clean water purging themselves of contaminants, including certain potentially pathogenic bacteria (Lee and Younger 2002). Overall, both studies showed that when applied to mussels, depuration decreases diversity of hepatopancreas microbiota. An increase in the *Vibrio* genus was observed in mussel collected in Italy, concurrently with the loss of rare microbial taxa, which was followed by the reorganization to a microbiome dominated by *Vibrio*. According to the authors, the dominance of this genus after depuration was a consequence of the appearance of new environmental niches during depuration that were occupied by the opportunistic *Vibrio*. In the case of mussel collected in Spain, depuration led to an increase of Tenericutes phylum along with decreases of Proteobacteria and Planctomycetes. The reduction observed for Proteobacteria was selective with a major impact on Gammaproteobacteria class, among which Moraxellaceae, Legionellaceae, and Enterobacteriaceae families suffered a higher impact than Vibrionaceae, more resilient to depuration. There was also a reduction in *Planctomycetes* and, as outlined above, an important increment of Tenericutes of unclear origin since this genus could not be identified from sequence data in undepurated samples. These bacteria have been involved in the pathogenesis of fish and invertebrates in some cases (Krol et al. 1991; Azevedo 1993) while being commensal in other cases (Boyle et al. 1987; Holben et al. 2002; Tanaka et al. 2004; Kellogg et al. 2009; Green and Barnes 2010).

Vibrio's association with mussel hepatopancreas can vary depending on the species occurring in the media. In a study carried out in the Greek coast, several potentially tetrodotoxin (TTX) producers, including *Vibrio alginolyticus* and *Vibrio parahaemolyticus,* were detected in mussels growing in TTX-contaminated waters when compared with control mussels from the same location (Rodríguez et al. 2017). This observation indicates that while *Vibrio* appears to be highly specific of mussel hepatopancreas, certain species can temporarily invade the organ under specific environmental conditions.

In summary, these studies support the stable presence of bacterial species with high affinity for mussel hepatopancreas that may even increase when the environmental conditions change. The functional importance of these bacteria for mussel's biology/physiology still needs to be worked out.

2.3 Shell Microbiota

The presence of high concentrations of mussel in restricted areas has been related to nitrogen cycling in the ocean (Lee and Childress 1995; Welsh and Castadelli 2004).

Later, the nitrogen metabolism observed was found to be related to microbial symbionts on the shell of these organisms. A study by shotgun metagenomic sequencing to analyze the shell microbiome of the California mussel (*M. californianus*) in the Tatoosh Island (Eastern Pacific) identified several major nitrifying bacterial genera (*Nitrosococcus*, *Nitrosomonas*, *Nitrospira*, *Nitrobacter*) in all the samples analyzed, as well as some denitrifying genera (*Shewanella and Roseobacter*) associated with a nitrogen-rich environment. Other bacteria also present on the shell included Cyanobacteria and Proteobacteria. In addition to the nitrogen cycling-related metabolism, a high microbial activity involved in carbon cycling metabolism was also observed. The aerobic pathway appears the most plausible for carbon cycling, considering the abundance of Cyanobacteria and Proteobacteria phyla and in particular the high concentration of Alpha-proteobacteria class on mussel surface (Pfister et al. 2010). The nitrogen cycling activity is responsible for the emission of N_2O, a powerful greenhouse gas, to the atmosphere. An analysis of N_2O production on *M. edulis* shell-associated biofilms showed that it contains a high denitrification microbial activity. In fact, the N_2O production of this species comes almost exclusively from the shell. The high production of N_2O on mussel surface can reach important levels in areas of high abundance like natural beds and intensive culture farms (Heisterkamp et al. 2013).

Another study on *M. californianus* from the same location analyzed mussel shell microbiome by 16S rDNA amplicon sequencing and shotgun metagenome sequencing. The taxonomic profiling showed a microbiome dominated by Gamma-proteobacteria but with a higher alpha diversity in the shell microbiota when compared to those of internal tissues. Operational taxonomic units (OTUs) from other phyla like Moritellaceae and Vibrionaceae, more specifically the genera *Moritella* and *Aliivibrio*, were also detected in different proportions. When compared with shotgun metagenomic data from the same study, the 16S rRNA gene amplicon sequences did not identify many taxa involved in nitrogen transformation, presumably as a result of the higher phylogenetic power of the shotgun sequencing, which enabled the identification of nitrogen transformation-associated genes from uncharacterized species. Even so, when the prediction of nitrogen metabolism present in the 16S rDNA sequenced samples was performed with PICRUST (Langille et al. 2013), several taxa involved in nitrogen metabolism were identified. This implies that metabolic inference from taxonomic analysis through 16S rDNA can sometimes provide similar outcomes to shotgun sequencing, as shown in this case (Pfister et al. 2014). If metagenome prediction is confirmed as a solid inference method in bacterial communities, it would represent an important step for the assessment of metabolic activities in different environments reducing the cost and effort of analyzing metagenomic data. This mostly relies on the available information in bacterial genome databases, which is increasing very rapidly thanks to the low cost of NGS. Anyway, metatranscriptomic analysis will be necessary to pinpoint active metabolism at a given condition in bacterial populations along with their taxonomic composition.

2.4 Gill Microbiota

Mussel feeding depends on the filtering of large water volumes from their living environments through their gills. In this scenario, the large number of bacteria ingested represents an important nutrient source after digestion, or may selectively end up on gills where they can potentially act as symbionts or commensals, being involved in metabolite synthesis and/or excretion, osmoregulation, etc. Due to this, gills represent an important target for microbiome studies to understand mussel physiology and production. To our knowledge, a single study has been addressed at analyzing mussel gill microbiome (Cappello et al. 2015). With the aim of identifying symbiotic relations between mussels and bacteria, the microbiome of gills was collected in Lake Faro (NE Sicilia, Italy) and analyzed by 16S rDNA and crDNA clone sequencing. While no clear symbiotic relationships were observed, the taxonomic composition of the microbiome (not complete due to the technique used and the number of clones sequenced) showed a high proportion of proteobacteria including the alpha, gamma, delta, and epsilon classes. Additionally, bacteria from the bacteroidetes, firmicutes, fusobacteria, actinobacteria, and chlamydiae phyla were identified. Interestingly, among all the samples sequenced only *Vibrio orientalis* was detected. The low proportion of *Vibrio* in this tissue could indicate that it is not retained in gills, an issue that still needs a deeper analysis taking into account the bacteriome of surrounding water and using a higher resolution technology like 16S rDNA amplicon sequencing or metagenomic analysis.

3 Biotic and Abiotic Factors That Control Mollusk Microbiota Assemblages

Adult mussels are considered robust and resistant to diseases (Venier et al. 2011; Philipp et al. 2012). Both a well-adapted immune system and a production system based on recruitment of natural seeds have been invoked to explain this phenomenon (Vera et al. 2011; Moreira et al. 2018). However, larvae suffer massive mortalities presumably due to microbial infections, a circumstance also observed in other bivalves (Beaz-Hidalgo et al. 2010; Genard et al. 2013). It has been speculated that there is a vertical transmission of bacteria from adult mussels to offspring among which some strain would become pathogenic, due to the physiological stress of the host organism (i.e., mussel larvae) in the context of high density growing (Eggermont et al. 2014). So, in addition to the immune system, it appears that microbiota might play an important role in the resistance to diseases of adults and the susceptibility of larvae. Accordingly, changes in bacterial populations can influence the susceptibility of mussel to infection throughout their life span. Several factors are capable of modulating mussel bacterial assemblages, many of which have not yet been addressed, and that could have notable effects on host physiology. In the next sections, the effects of abiotic and biotic factors on the mussel microbiota are reviewed.

3.1 Abiotic Factors

Mussel bacterial assemblages are susceptible to important changes due to various environmental factors like pH, temperature, and dissolved oxygen. Changes in these factors may have a direct effect on the living microbiota by altering the diversity and structure of the bacterial community, which, in turn, will modify the ability of the mussel to convert food into energy. These bacteria not only colonize the body surface and digestive tract of mussels but also can enter into the hemolymph. This environmentally modified microbiota could eventually affect mussel growth, health, and/or reproduction.

The decrease of the ocean's pH is expected to be around 0.2 units this century as a consequence of the human-induced increase of atmospheric CO_2, and it could have a substantial impact on marine ecosystems due to its critical role in mediating physiological reactions (Wootton et al. 2008). It has been speculated that prolonged exposure to pH shifts could produce significant changes in the microbial composition of the oceans. Experimental data have shown that a small pH shift has determined a significant variation in the microbial community of the North Sea (Krause et al. 2012) and Ross Sea (Antarctica) (Maas et al. 2013). An example of the potential effect of CO_2 concentration on mussels may be detected by changes in the tightly associated *Vibrio* genus. As indicated in the previous section, this genus appears closely associated with mussels, being present not only in the digestive tract but also on the shell and in the hemolymph. As highlighted before, mussel health depends on its tolerance to *Vibrio* to avoid bacterial diseases, and therefore on the growth conditions and concentration of *Vibrio* in the surrounding waters, which can turn commensal Vibrios into pathogenic. Reduced pH has been shown to favor the growth of *Vibrionaceae* (Meron et al. 2011), and consequently, favorable conditions for the growth of this genus might negatively affect mussel health. Populations would therefore be exposed to pathogenic Vibrios or to an increase of commensal Vibrios to a degree that they might become pathogenic. Also, an impact on biofilms would be expected through a shift from bacterial to algal communities dominance with the consequent effects on productivity (Rost et al. 2008). This impact has been investigated in the Australian Great Barrier Reef where an increase in Flavobacteriales along with a decrease in Alpha-proteobacteria has been observed in biofilms as a consequence of lower pH, induced by elevated atmospheric CO_2 (Witt et al. 2011). The authors of this work hypothesized that biofilm alteration could have detrimental effect on corals reefs since, as primary reef colonizers, biofilms allow the settlement and development of different invertebrates. While there is no published evidence of changes on mussel shell microbiota in response to decreasing pH, it is reasonable to speculate that shell microbiota might suffer important changes in composition as a consequence of decreasing ocean's pH according to that observed in corals. As in corals, shell microbiota of mussels has high proportions of Alpha-proteobacteria and a certain amount of Flavobacteriales. The effect of lowering pH on mussel gut, hemolymph, and gill microbiota, as well as the consequence of these bacterial population shifts on mussel growth, health, and proliferation, still needs to be addressed.

The decline in ocean's pH will probably also affect the mechanisms of bacterial communication (quorum sensing), a key process that influences microbial interaction in all trophic levels of the ecosystem. This process depends on autoinducing peptides (AIPs) in Gram-positive bacteria and on acyl homoserine lactones (AHLs) in Gram-negative bacteria. Both have important biological and ecological functions regulating symbiosis, competence, virulence, secondary metabolite production, extracellular enzymes, biofilm formation, and bioluminescence in the marine environment (Miller and Bassler 2001; Manefield and Whiteley 2007; De Kievit 2009; Weber et al. 2009; Chong et al. 2012; Mangwani et al. 2012). Proteobacteria are among the main AHL producers in oceans (Manefield and Turner 2002). Again, corals represent an example of the effect of ocean acidification on quorum sensing. It has been shown that acidification favors the growth of pathogenic bacteria in the water surrounding coral reefs, which triggered the synthesis of secondary metabolites as a protective mechanism by the coral itself and its associated microorganisms (Skindersoe et al. 2008; Golberg et al. 2011). To what extent ocean acidification will impact on mussel-associated microbiota at different levels is still a matter of debate.

The effect of climate change on ocean's microbial populations is hard to predict. Even so, some evidence points toward an increase of pathogenic species as a consequence of increased temperature on ocean's surface (Vezzulli et al. 2012), putatively a few degrees in this century (Harvell et al. 2002). The effect of global warming has been linked to higher incidence of *Vibrio*-related diseases. This increase has been observed not only in humans (Martinez-Urtaza et al. 2010) but also in the coastal marine species including mussels (Paillard et al. 2004; Vezzulli et al. 2010). While the higher *Vibrio* incidence in pathogenic episodes is not necessarily related to the proliferation of bacteria of this genus in coastal marine waters, experiments carried out with plankton provide evidence on this regard. Using samples from an historical archive spanning 44 years, Vezzulli et al. (2012) showed the increasing dominance of Vibrios, including *Vibrio cholerae*, significantly correlated with warming in the sea surface during the same period. In fact, the major change in bacterial community composition was that of the Vibrionaceae family (Vezzulli et al. 2012). As outlined before, Vibrios are commonly associated with hepatopancreas, hemolymph, and shell of mussels (Olafsen et al. 1993; Antunes et al. 2010; Pfister et al. 2014; Rodríguez et al. 2017; Vezzulli et al. 2018; Rubiolo et al. 2018), and in fact, when present, pathogenic *Vibrio* species are accumulated in mussel gut displacing other bacteria (Rodríguez et al. 2017). While being commensal bacteria in normal conditions, they can become pathogenic under stress or under conditions that promote pathogenic species growth in the mussel surrounding waters triggering microbiota dysbiosis and consequently the weakening of colonization resistance proprietary of the established microbiota toward potential invaders. Accordingly, the increased proportion of pathogenic *Vibrio* spp., a consequence of increased sea surface temperature, may result in higher mortalities and reduced growth of mussels. Considering that such a short period of water warming has been enough to observe significant changes in the microbiota composition in surface seawater, other unpredictable changes would emerge if this trend increases.

3.2 Biotic Factors

Changes in the biotic factors affecting mussel microbiota most likely derive from anthropogenic changes related to climate change (see previous section) or nutrient enrichment and from periodic natural events like El Niño current and associated upwelling. The mussel ecosystem is then affected, not only on its physical parameters as previously outlined, but also by changing the microbial composition of seawater, which in turn affects the mussel microbiota at different levels. There is ample evidence showing that nonindigenous bacteria can colonize mussel gut displacing indigenous microorganisms (Rodríguez et al. 2017; Vezzulli et al. 2018; Rubiolo et al. 2018) and also that indigenous microbiota in certain conditions can become pathogenic. In this regard, Eggermont et al. (2014) have shown how increasing nutrients can stimulate mussel indigenous heterotrophic bacteria resulting in mussel mortality. These authors, using a controlled environment where only mussel endogenous bacteria were present, demonstrated that increasing the amount of nutrients in the water promoted the proliferation of heterotrophic bacteria, which resulted in massive mortality of *Mytilus edulis*. These results suggest that opportunistic pathogens, which in normal conditions are associated with adult mussels, can become pathogenic when the conditions are favorable for their proliferation. The vertical transmission of these bacteria to larvae could be at least one of the reasons of massive mussel larvae mortality when cultured at high densities, where higher load of organic matter is present in surrounding water (Eggermont et al. 2014). Further, 17 bacterial strains, which induced a wide range of mortality in *Mytilus edulis* larvae (17–98%), were isolated using a similar experimental design. The strains identified were from the Splendidus clade of *Vibrio*, previously associated with mass mortalities in bivalve larvae (Sugumar et al. 1998; Kesarcodi-Watson et al. 2009; Kwan and Bolch 2015; Rojas et al. 2015; Travers et al. 2015), and *Photobacterium*. Whole genome sequencing allowed the identification of *Vibrio hemicentroti* and *Photobacterium sanguinicancri* as the most virulent species, both harboring several genes involved in virulence processes in bivalves, such as proteases and hemolysins. Besides, all other isolates expressed similar virulence factors and showed antibiotic resistance (Eggermont et al. 2017). All in all, these results indicate that wild mussels house a reservoir of potentially pathogenic bacteria that, under specific environmental conditions, can produce important mortalities to themselves. While these experiments were performed in laboratory-controlled conditions, it is likely that anthropogenic coastal eutrophication may create the conditions necessary for growth of opportunistic pathogens for mussels resulting in reduced viability. In fact, a significant association between 27 bacterial families, including Gamma-proteobacteria, the phylum containing *Vibrio* spp., among others, and the eutrophication index has been observed (Dai et al. 2017).

4 Tools for Modulating Water Community and Mussel Microbiota to Improve Viability, Health, and Growth Performance

It is accepted that symbiotic microbiota can provide protection to the host by avoiding or limiting the settlement of pathogens through competing for resources, immune system stimulation, and/or production of secondary metabolites (antibiotics) (Oelschlaeger 2010). Thus, it makes the mussel's microbiota a reservoir of beneficial microbial functions that can be harnessed to enable the control of the commensal and/or pathogenic microbial community (Dobson et al. 2012). As explained in the previous sections of this chapter, several environmental biotic and abiotic pressures exist, and new ones are expected to emerge, for example, as a consequence of the climate change. In this context, indigenous and nonindigenous microorganisms (probiotics) capable of competing with pathogens might serve as a useful tool to protect the host (Desriac et al. 2010).

Despite the few mass mortalities documented (Genard et al. 2013; Ben Cheikh et al. 2016), adult mussels are considered robust and not prone to infections. This is not the case of larvae, which, as outlined above, are susceptible to high rates of mortality in dense cultures (Sainz-Hernández and Maeda-Martínez 2005). The mussel industry depends on natural spat, which has sustained the activity for many years. Spat availability is subject to natural and anthropogenic pressures like temperature shifts and nutrient availability that can have a direct impact on their survival or affect predators and emergent pathogens. An alternative to spat collection from nature is the hatchery production of seed (FAO 2004), which has some advantages like the possibility of establishing selection programs (Pino-Querido et al. 2015). This technology, despite available, is not widely applied due to the high costs associated with feed (microalgae) production, the low market value of mussel when compared to other bivalves, and, as previously outlined, the mass mortalities occurring in dense larval cultures (FAO 2004; Sainz-Hernández and Maeda-Martínez 2005). Mass mortality in hatcheries, where much larger amount of nutrients is reached, is probably a consequence of microbial infections (Powell et al. 2013). The higher load of organic matter can facilitate the growth of opportunistic pathogens in the rearing water massively affecting larval culture. Several sanitary strategies have been tested to eliminate the microbiota of the rearing seawater such as water treatment or chemotherapy but with limited success. Probiotics have emerged as a promising alternative to solve this problem.

4.1 Rearing Conditions

In bivalve hatcheries, where larvae share the environment with microbiota containing both beneficial and pathogenic bacteria, there are two main methods that have been applied to deal with infections: water treatment and chemotherapy.

This is essential since larvae are especially sensitive to potential infections and are hosted in a closed environment with a regular supply of food containing bacteria. To reduce the bacterial biomass, different treatments can be applied including filtration, ozonization, chlorine disinfection, or UV radiation. Filtration reduces the heterotrophic bacterial population in the water. A regrowth of these bacteria can be later observed as a result of the daily phytoplankton supply (Jeanthon et al. 1988). Ozonization has several disadvantages including the cost and the oxidants produced that are toxic not only to aquaculture species but also to humans (Richardson et al. 1982; Summerfelt 2003). These drawbacks together with the fact that its effectiveness has not been sufficiently studied make this method inappropriate for bivalve hatcheries. Chlorine disinfection of water presents problems too, since it interferes with the pumping mechanism of larvae and produces toxic residues by reacting with organic nitrogen (Vasconcelos and Lee 1972; Jorquera et al. 2001). The most usual treatment is UV irradiation, which has been tested by different authors showing quite different outcomes. A reduction of the populations of *Pseudomonas*, *Vibrios*, Coliforms, and Gram-positive cocci with increased *Acinetobacter* and *Moraxella* was observed on larval cultures by Vasconcelos and Lee (1972). Conversely, other authors recorded decreased concentrations of *Vibrio*, *Achromobacter*, and *Flavobacterium* along with increased *Pseudomonas* (Murchelano et al. 1975). In both assays, the UV treatment reduced the diversity of the microbiota present in the water. Due to its relevance for bivalves' larvae pathogenesis and mass mortalities, the effect of UV radiation on *Vibrio* has been studied more thoroughly. Several studies showed that even when the population of Vibrios decreased after water irradiation, the effect was not homogeneous, affecting some species more than others (Brown 1981; Lodeiros et al. 1987). Furthermore, other important factors such as dose, water flow, water organic content, and individual efficiency of the radiation unit have been documented. In some cases, this treatment is more bacteriostatic than bactericidal producing only a temporal inhibition of bacterial growth (Brown 1983; Liltved et al. 1995; Liltved and Cripps 1999). The efficiency reached with this technique is rather variable depending on various factors and can lead to the enrichment of certain bacterial populations, which, considering the high cost of irradiation, makes its use not so straightforward or affordable for commercial production. This puts an even greater burden on mussel production, which limits its applicability due to its low commercial value when compared to other bivalves.

Another way to control the bacterial populations in bivalve's hatcheries is the use of antibiotics. Initial studies using different antibiotic combinations in Mediterranean mussel showed that chloramphenicol increased survival rate but reduced the larvae development. On the other hand, the ampicillin–streptomycin combination enabled a good development but with lower survival rate (Hily 1974). Antibiotic therapies have been assayed in other bivalves with different outcomes. For example in oysters, chloramphenicol had a similar effect to that observed in mussels, while erythromycin was toxic at the effective dose (Jeffries 1982; Brown and Tettelbach 1988). Chloramphenicol showed inconsistent results in the scallop *Pecten maximus* (Robert et al. 1996), and rifampicin increased survival but produced malformations in the giant clam *Tridacna derasa* (Fitt et al. 1992).

Several objections have been raised to antibiotic therapy, the appearance of resistances being likely the most important. Its use as preventive treatment in hatcheries has potential harmful effects on human health, but it can also be toxic to the invertebrate fauna. For example, chloramphenicol, the most used antibiotic in bivalve hatcheries, is extremely toxic. It may produce larval malformations, favors the development of resistances, and in addition, is harmful to humans. Moreover, its high cost of production with low profit margins, such as that for mussels, can mean an unsustainable economic burden (Baticados et al. 1990). The aforementioned treatments (water treatment by filtration or UV irradiation, or antibiotic therapy) are applied to obtain the full elimination of the microbiota associated with rearing water. While certain members of this microbiota could represent a threat to reared larvae, others could be beneficial, for example as nutrient source, thus satisfying metabolic requirements and/or preventing growth of potential pathogens (Prieur et al. 1990). Also, pathogen growth could be augmented by the complete elimination of microbiota. In a context of high organic matter supply and the absence of competing bacteria, the incoming bacteria associated with food and larvae could favor the fast growth of the opportunistic pathogen *Vibrio* spp. as previously outlined. These facts along with the lack of cost-effective solutions for the control of bacterial diseases in bivalve larval cultures highlight the crucial need for new solutions; this is where probiotics come into play.

4.2 Food Additives (Probiotics)

Elie Metchnikoff was the first to describe the positive role of some bacteria on human health identifying the possibility of replacing harmful microbes by beneficial ones (Metchnikoff 1908). Probiotics ("food for life" from the Greek *pro=for* and *bios=life*), a word introduced in 1965 as a modification of the word probiotika (Lilly and Stillwell 1965), were used to define the effect of living nonpathogenic organisms and their beneficial effects on hosts. Based on the first studies performed in homeotherms, mostly humans and rodents, probiotics were initially defined as "non-pathogenic microorganisms which when ingested, exert a positive influence on host's health or physiology" (Fuller 1989), and later, as "living microbial cells administered as dietary supplements with the aim of improving health" (Tannock 1997). The latest FDA and WHO definition is "live microorganisms which when administered in adequate amounts confer a health benefit to the host." These definitions were later refined by several authors in order to include organisms in aquaculture and eliminate its restriction to the intestinal balance. The most recent definition that probably better applies to bivalve larval culture, among others that have been proposed, and which is still a matter of controversy (Gatesoupe 1999; Irianto and Austin 2002; Balcazar et al. 2006) states that a probiotic is a "live microbial adjunct which has a beneficial effect on the host by modifying the host-associated or ambient microbial community, by ensuring improved use of the feed or enhancing its nutritional value, by enhancing the host response towards disease, or by improving the quality of its ambient environment" (Verschuere et al. 2000). The beneficial effects of probiotics on bivalves are variable and include improvement of water quality, better use of feed, nutritional complement, antibacterial activity, adhesion, or metamorphosis. Probiotic products can be composed of a single or a mixture of two or more strains. The probiotic effect is strain-specific and a single strain can exhibit different benefits when used individually or in combination (Chapman et al. 2011). Some characteristics of probiotic organisms adapted to aquaculture are the following:

- Nonpathogenic
- Tolerant to acids
- Short regeneration time
- Robust and surviving processing conditions
- Anti-genotoxic
- Genetically stable
- Lack of resistance to antibiotics

4.3 Probiotics Intervention in Mussel in the Context of Mollusk Culture

There are several reports on the use and performance of probiotics in bivalve species (Table 1). In all cases, antibacterial activity with increased bivalve survival rate was observed. Most of the probiotic bacteria identified belonged to Gamma-proteobacteria like *Alteromonas* sp. CA2, *Pseudomonas* sp. 11, *Vibrio* sp. 13, *Aeromonas* media A199, and *Pseudoalteromonas* spp. These strains have shown effective antibacterial activity and/or have enhanced growth rate in *Crassostrea gigas*, *Argopecten purpuratus*, *Pecten maximus*, *Perumytilus purpuratus*, *Crassostrea virginica*, *Argopecten irradians*, and *Ostrea edulis* larval cultures (Douillet and Langdon 1993, 1994, Riquelme et al. 1996, 1997, 2001; Gibson et al. 1998; Avendaño and Riquelme 1999; Longeon et al. 2004; Aranda et al. 2012, Karim et al. 2013; Sohn et al. 2016). Besides Gamma-proteobacteria, members of other clades have demonstrated a probiotic action in bivalve larval cultures such as *Roseobacter gallaeciensis* and *Phaeobacter gallaeciensis* (Ruiz-Ponte et al. 1999; Prado et al. 2009). The antibacterial effects of the identified probiotic strains are effective against a wide range of Vibrios, *E. coli*, *A. salmonicida*, *S. putrefaciens*, and *S. aureus*, among others (for a thorough review, see Prado et al. 2010). These results support the use of probiotics for controlling the microbiota at bivalve hatcheries.

As shown in Sect. 2.1, mussel hemolymph is not sterile and contains microorganisms indicating that a particular equilibrium exists between the host and these microorganisms. Based on the idea that the bacterial hemolymph might provide host resistance to environmental pressures, the hemolymph cultivable bacteria of several bivalves have been investigated for their probiotic activity, testing their antibacterial potential. The hemolymph of Mediterranean mussel showed a higher bacterial

Table 1 Probiotics with positive effects on survival rate of bivalve larvae

Probiotic strain[a]	Challenge bacteria[b]	Effect	Bivalve species	Authors
Flavobacterium sp. P14	*Vibrio anguillarum* EPp3	Antibiotic	*Pecten ziczac*	Lodeiros et al. (1989)
Alteromonas sp. CA2	–	Increased growth and survival rate	*Crassostrea gigas*	Douillet and Langdon (1993, 1994)
Pseudoalteromonas haloplanktis	*Vibrio anguillarum*	Antibiotic	*Argopecten purpuratus*	Riquelme et al. (1996)
Pseudomonas sp. 11 *Vibrio* sp. C33	*Vibrio anguillarum*-related	Antibiotic	*Argopecten purpuratus*	Riquelme et al. (1997)
Aeromonas media A199	*Vibrio tubiashii*	Antibiotic	*Crassostrea gigas*	Gibson et al. (1998)
Vibrio sp. C33	*Vibrio anguillarum*-related	Antibiotic	*Argopecten purpuratus*	Avendaño and Riquelme (1999)
S21	*Vibrio alginolyticus*	Antibiotic	*Crassostrea gigas*	Nakamura et al. (1999)
Phaeobacter gallaeciensis BS107	*Vibrio pectenicida*	Antibiotic	*Pecten maximus*	Ruiz-Ponte et al. (1999)
Vibrio sp. C33 *Pseudomonas* sp. 11 *Bacillus* sp. B2	–	Increased survival rate	*Argopecten purpuratus*	Jorquera et al. (2001)
Pseudoalteromonas sp.	–	Increased survival rate	*Pecten maximus*	Longeon et al. (2004)
Isolates 0536, 0548, 0599, 04287, 0532, 0594, 0448, 0593	*Vibrio* sp. DO1 *Vibrio splendidus*	Increased survival rate	*Perna canaliculus*	Kesarcodi-Watson et al. (2009)
Isolate OY15	*Vibrio coralliilyticus*-like B183	Antibiotic	*Crassostrea virginica*	Lim et al. (2011)
Alteromonas macleodii 0444 *Neptunomonas* sp. 0536, *Phaeobacter gallaeciensis*, *Pseudoalteromonas* sp. D41	*Vibrio coralliilyticus* *Vibrio splendidus*	Increased survival rate	*Ostrea edulis* *Pecten maximus* *Crassostrea gigas*	Kesarcodi-Watson et al. (2012)
Phaeobacter sp. S4 *Bacillus pumilus* RI06-95	*Roseovarius crassostreae* CV919-312[T] *Vibrio coralliilyticus* RE22	Antibiotic Increased survival rate	*Crassostrea virginica*	Karim et al. (2013)
Phaeobacter inhibens S4 *Bacillus pumilus* RI06–95	*Vibrio coralliilyticus* RE22	Increased survival rate	*Argopecten irradians*	Sohn et al. (2016)

[a]Isolates, as named in the original works cited, are indicated in this column for probiotics with no gender assigned
[b]When necessary bacteria gender and species have been renamed to actual accepted classification

concentration when compared to *T. rhomboides*, *P. maximus*, and *C. gigas*. Of the 843 bacterial isolates obtained from all bivalves analyzed, 26 showed antibacterial activity. 16S rDNA sequencing identified these bacteria as belonging to the genus *Pseudoalteromonas*, *Thalassomonas*, and *Vibrio*. Among these, 14 *Pseudoalteromonas*, two *Vibrio*, and one not identifiable isolate were detected in Mediterranean mussel. These bacteria showed antibacterial effect against known bivalves *Vibrio* pathogens (*V. splendidus*, *V. tapetis*), and three *Pseudoalteromonas* also showed antibacterial effect against other aquaculture pathogens, including an effective strain against *Listonella anguillarum*, *Aeromonas caviae*, *Aeromonas hydrophila*, and *Aeromonas salmonicida*. The two other strains were effective against only three of these pathogens (Desriac et al. 2014). The identification of strains exhibiting potential probiotic properties shows that mussel hemolymph along with that of other bivalves could provide new antimicrobial tools adding new evidences to the hypothesis that certain bacteria present in the hemolymph may protect the host through their antimicrobial effect.

4.4 First Evidences of Effective Interventions in Mussel Larval Production

The continuous supply of spat is crucial for the bivalve aquaculture industry. Disease outbreaks in hatcheries can greatly affect larval supply, and accordingly, the management of bacterial diseases is a key to warrant sustainable production. For this purpose, several technologies such as water treatment and/or antibiotic therapy are applied, but they have important drawbacks (see Sect. 4.1). While probiotics have been assayed in several bivalves, most notably in oysters, very few research efforts have been done on mussels. Even though there are initial and promising results, further deeper analyses will be necessary. Efforts have been made to identify probiotics that prevent larvae mortalities in green shell mussels (*Perna canaliculus*) by bacterial pathogens in hatcheries. This work resulted in 69 bacterial isolates, 40 of them showing a significant protective effect against *V. splendidus*, apparently by competitive gut colonization (possibly by preventing pathogen attachment) and posterior infection and multiplication. These results, initially obtained in laboratory conditions, were further tested in hatchery for a reduced set of candidate probiotics. The effect of these candidates was evaluated in larvae challenged with two pathogens, *V. splendidus* and *Vibrio* spp. DO1. Among ten potential probiotics assayed, eight significantly reduced the mortality induced by *Vibrio* spp. DO1, while three showed the same effect against *V. splendidus* (Kesarcodi-Watson et al. 2009). This pilot study in hatchery conditions shows that probiotic treatment could be effective for the prevention of mass mortalities in mussel larvae in hatcheries. A more recent work analyzed the effect of the probiotic strains *Phaeobacter inhibens* S4 and *Bacillus pumilus* RI06–95, which had been successfully applied to prevent oyster larvae mortality by bacterial infection (Karim et al. 2013), for their potential against the pathogen *Vibrio coralliilyticus* on several mollusks including blue mussel, northern quahogs (*Mercenaria mercenaria*), bay scallops (*Argopecten irradians*), and razor clams (*Ensis directus*). All the bivalve species assayed were susceptible to the tested pathogen, but when pretreated with the probiotics, oysters and bay scallop exhibited a higher survival rate relatively to controls. On the contrary, no beneficial effect was observed in mussel larvae and even a detrimental one was observed in razor clams, showing a higher susceptibility to the bacterial challenge (Sohn et al. 2016). These results suggest that, despite information from other species providing interesting candidates, the isolation of endogenous probiotics will probably end up identifying more effective bacterial strains.

5 Research Gaps and Future Perspectives

Mussels are the most important aquaculture mollusks in Europe (>300,000 tons). Nonetheless, when compared to other bivalves, they have attracted little attention for the study of their associated microbiome and the development of useful probiotics for production. This is likely due to the robust nature of adult mussels regarding bacterial infections, its low market value, and the fact that currently spats are mostly collected in the wild. However, the situation could change soon, due to the combined effect of several anthropogenic changes, including global climate change and eutrophication, which could change the bacterial communities to which mussels are adapted. There is a complex equilibrium between mussels and its associated microbiome implying that environmental changes can turn commensal bacteria into pathogenic ones. Therefore, mussels are not only facing the exposure to new nonindigenous bacteria but also to the alteration of their own microbiota composition. While the consequences of these changes are hard to be predicted, identifying potential pathogens and/or detrimental bacteria through 16S metabarcoding and shotgun metagenomics studies in mussels and surrounding waters in relevant geographic locations could shed some light on their effects. Environmental changes are also putting pressure on the production of these bivalves since the availability of seed needed for commercial production might be at risk.

Since mussel spat is still available for collection in nature, hatchery production has hardly been developed. Due to shortages in spat availability and the previously aforementioned changes that could increase outbreak episodes, the production of larvae in hatcheries is becoming a key factor for production. Mass mortalities as a result of bacterial infections are the main problem for this activity, so it is imperative to develop tools that can deal with this threat. Since water treatment and antibiotic therapy have been shown to be far from ideal, in addition to the fact that the use of antibiotics is no more allowed in many cases, the development of probiotics appears as a promising alternative. While several bacterial strains have been identified for use in many aquaculture species including bivalves, scarce information is available in mussels. The studies compiled in this review show that several bacterial strains have a probiotic effect in laboratory-controlled conditions, and pilot studies in hatcheries suggest that probiotic treatments of larvae might be efficient. Further effort should be undertaken toward the identification, test, and validation of probiotic strains in the context of industrial production.

References

Alber M, Valiela I (1994) Production of microbial organic aggregates from macrophyte-derived dissolved organic material. Limnol Oceanogr 39:37–50. https://doi.org/10.4319/lo.1994.39.1.0037

Antunes F, Hinzmann M, Lopes-Lima M et al (2010) Association between environmental microbiota and indigenous bacteria found in hemolymph, extrapallial fluid and mucus of Anodonta cygnea (Linnaeus, 1758). Microb Ecol 60:304–309. https://doi.org/10.1007/s00248-010-9649-y

Aranda CP, Valenzuela C, Barrientos J et al (2012) Bacteriostatic anti-Vibrio parahaemolyticus activity of Pseudoalteromonas sp. strains DIT09, DIT44 and DIT46 isolated from Southern Chilean intertidal Perumytilus purpuratus. World J Microbiol Biotechnol 28:2365–2374. https://doi.org/10.1007/s11274-012-1044-z

Avendaño RE, Riquelme CE (1999) Establishment of mixed-culture probiotics and microalgae as food for bivalve larvae. Aquacult Res 30:893–900. https://doi.org/10.1046/j.1365-2109.1999.00420.x

Azevedo C (1993) Occurrence of an unusual branchial mycoplasma-like infection in cockle Cerastoderma edule (Mollusca, Bivalvia). Dis Aquat Organ 16:55–59. https://doi.org/10.3354/dao016055

Balbi T, Fabbri R, Cortese K et al (2013) Interactions between Mytilus galloprovincialis hemocytes and the bivalve pathogens Vibrio aestuarianus 01/032 and Vibrio splendidus LGP32. Fish Shellfish Immunol 35:1906–1915. https://doi.org/10.1016/j.fsi.2013.09.027

Balcazar JL, Blas I, Ruiz-Zarzuela I et al (2006) The role of probiotics in aquaculture. Vet Microbiol 114:173–186

Baticados MCL, Lavilla-Pitogo CR, Cruz-Lacierda ER et al (1990) Studies on the chemical control of luminous bacteria Vibrio harveyi and V. splendidus isolated from diseased Penaeus monodon larvae and rearing water. Dis Aquat Organ 9:133–139

Beaz-Hidalgo R, Balboa S, Romalde JL, Figueras MJ (2010) Diversity and pathogenecity of Vibrio species in cultured bivalve molluscs. Environ Microbiol Rep 2:34–43

Beleneva I, Zhukova N, Maslennikova E (2003) Comparative study of microbial communities from cultured and natural populations of the mussel Mytilus trossulus in Peter the Great Bay. Microbiology 72:472–477

Ben Cheikh Y, Travers MA, Morga B et al (2016) First evidence for a Vibrio strain pathogenic to Mytilus edulis altering hemocyte immune capacities. Dev Comp Immunol 57:107–119. https://doi.org/10.1016/j.dci.2015.12.014

Boyle PJ, Maki JS, Mitchell R (1987) Mollicute identified in novel association with aquatic invertebrate. Curr Microbiol 15:85–89. https://doi.org/10.1007/BF01589367

Brown C (1981) A study of two shellfish-pathogenic Vibrio strains isolated from a Long Island hatchery during a recent outbreak of disease. J Shellfish Res 1:83–87

Brown C (1983) Bacterial diseases in bivalve larval cultures and their control. In: Berg C Jr (ed) Culture of marine invertebrates. Hutchinson Ross, Stroudsburg, PA, pp 230–242

Brown C, Tettelbach LP (1988) Characterization of a nonmotile Vibrio sp. pathogenic to larvae of Mercenaria mercenaria and Crassostrea virginica. Aquaculture 74:195–204. https://doi.org/10.1016/0044-8486(88)90363-8

Canesi L, Gallo G, Gavioli M, Pruzzo C (2002) Bacteria-hemocyte interactions and phagocytosis in marine bivalves. Microsc Res Tech 57:469–476. https://doi.org/10.1002/jemt.10100

Canesi L, Grande C, Pezzati E et al (2016) Killing of Vibrio cholerae and Escherichia coli strains carrying D-mannose-sensitive ligands by Mytilus hemocytes is promoted by a multifunctional hemolymph serum protein. Microb Ecol 72:759–762. https://doi.org/10.1007/s00248-016-0757-1

Cappello S, Volta A, Santisi S et al (2015) Study of bacterial communities in mussel Mytilus galloprovincialis (Bivalvia: Mytilidae) by a combination of 16s Crdna and 16s Rdna sequencing. JSM Microbiol 3:1016

Carballal M, Lopez C, Azevedo C, Villalba A (1997) Enzymes involved in defense functions of hemocytes of mussel Mytilus galloprovincialis. J Invertebr Pathol 70:96–105. https://doi.org/10.1006/jipa.1997.4670

Cavallo RA, Acquaviva MI, Stabili L (2009) Culturable heterotrophic bacteria in seawater and Mytilus galloprovincialis from a Mediterranean area (Northern Ionian Sea—Italy). Environ Monit Assess 149:465–475. https://doi.org/10.1007/s10661-008-0223-8

Chapman CMC, Gibson GR, Rowland I (2011) Health benefits of probiotics: are mixtures more effective than single strains? Eur J Nutr 50:1–17

Cheng TC (1975) Functional morphology and biochemistry of molluscan phagocytes. Ann N Y Acad Sci 266:343–379. https://doi.org/10.1111/j.1749-6632.1975.tb35116.x

Chong G, Kimyon O, Rice SA et al (2012) The presence and role of bacterial quorum sensing in activated sludge. Microb Biotechnol 5:621–633. https://doi.org/10.1111/j.1751-7915.2012.00348.x

Cole J, Findlay S, Pace M (1988) Bacterial production in fresh and saltwater ecosystems: a cross-system overview. Mar Ecol Prog Ser 43:1–10. https://doi.org/10.3354/meps043001

Collin B, Rehnstam-Holm AS (2011) Occurrence and potential pathogenesis of Vibrio cholerae, Vibrio parahaemolyticus and Vibrio vulnificus on the South Coast of Sweden. FEMS Microbiol Ecol 78:306–313. https://doi.org/10.1111/j.1574-6941.2011.01157.x

Craft JA, Gilbert JA, Temperton B et al (2010) Pyrosequencing of Mytilus galloprovincialis cDNAs: tissue-specific expression patterns. PLoS One. https://doi.org/10.1371/journal.pone.0008875

Dai W, Zhang J, Tu Q et al (2017) Bacterioplankton assembly and interspecies interaction indicating increasing coastal eutrophication. Chemosphere 177:317–325. https://doi.org/10.1016/j.chemosphere.2017.03.034

De Kievit TR (2009) Quorum sensing in Pseudomonas aeruginosa biofilms. Environ Microbiol 11:279–288

Desriac F, Defer D, Bourgougnon N et al (2010) Bacteriocin as weapons in the marine animal-associated bacteria warfare: inventory and potential applications as an aquaculture probiotic. Mar Drugs 8:1153–1177

Desriac F, Le Chevalier P, Brillet B et al (2014) Exploring the hologenome concept in marine bivalvia: haemolymph microbiota as a pertinent source of probiotics for aquaculture. FEMS Microbiol Lett 350:107–116

Dobson A, Cotter PD, Paul Ross R, Hill C (2012) Bacteriocin production: a probiotic trait? Appl Environ Microbiol 78:1–6

Douillet P, Langdon CJ (1993) Effects of marine bacteria on the culture of axenic oyster Crassostrea gigas (Thunberg) Larvae. Biol Bull 184:36. https://doi.org/10.2307/1542378

Douillet PA, Langdon CJ (1994) Use of a probiotic for the culture of larvae of the Pacific oyster (Crassostrea gigas Thunberg). Aquaculture 119:25–40. https://doi.org/10.1016/0044-8486(94)90441-3

Dubert J, Barja JL, Romalde JL (2017) New insights into pathogenic vibrios affecting bivalves in hatcheries: present and future prospects. Front Microbiol 8:762. https://doi.org/10.3389/fmicb.2017.00762

Eggermont M, Tamanji A, Nevejan N et al (2014) Stimulation of heterotrophic bacteria associated with wild-caught blue mussel (Mytilus edulis) adults results in mass mortality. Aquaculture 431:136–138

Eggermont M, Bossier P, Pande GSJ et al (2017) Isolation of Vibrionaceae from wild blue mussel (Mytilus edulis) adults and their impact on blue mussel larviculture. FEMS Microbiol Ecol 93 (4). https://doi.org/10.1093/femsec/fix039

Elston R, Leibovitz L, Relyea D, Zatila J (1981) Diagnosis of vibriosis in a commercial oyster hatchery epizootic: diagnostic tools and management features. Aquaculture 24:53–62. https://doi.org/10.1016/0044-8486(81)90043-0

Engel S, Jensen PR, Fenical W (2002) Chemical ecology of marine microbial defense. J Chem Ecol 28:1971–1985. https://doi.org/10.1023/A:1020793726898

FAO (2004) Cultured aquatic species information programme. Mytilus edulis (Linnaeus, 1758). Rome

Fitt WK, Heslinga GA, Watson TC (1992) Use of antibiotics in the mariculture of giant clams (F. Tridacnidae). Aquaculture 104:1–10. https://doi.org/10.1016/0044-8486(92)90132-5

Fuller R (1989) Probiotics in man and animals. J Appl Bacteriol 66:365–378

Gatesoupe FJ (1997) Siderophore production and probiotic effect of Vibrio sp. associated with turbot larvae, *Scophthalmus maximus*. Aquat. Living Resour 10:239–246

Gatesoupe FJ (1999) The use of probiotic in aquaculture. Aquaculture 180:147–165

Genard B, Miner P, Nicolas J-L et al (2013) Integrative study of physiological changes associated with bacterial infection in Pacific oyster larvae. PLoS One 8:e64534. https://doi.org/10.1371/journal.pone.0064534

Gibson LF, Woodworth J, George AM (1998) Probiotic activity of Aeromonas media on the Pacific oyster, Crassostrea gigas, when challenged with Vibrio tubiashii. Aquaculture 169:111–120. https://doi.org/10.1016/S0044-8486(98)00369-X

Golberg K, Eltzov E, Shnit-Orland M et al (2011) Characterization of quorum sensing signals in coral-associated bacteria. Microb Ecol 61:783–792. https://doi.org/10.1007/s00248-011-9848-1

Green TJ, Barnes AC (2010) Bacterial diversity of the digestive gland of Sydney rock oysters, Saccostrea glomerata infected with the paramyxean parasite, Marteilia sydneyi. J Appl Microbiol 109:613–622. https://doi.org/10.1111/j.1365-2672.2010.04687.x

Grimes DJ, Colwell RR, Stemmler J et al (1984) Vibrio species as agents of elasmobranch disease. Helgoländer Meeresuntersuchungen 37:309–315. https://doi.org/10.1007/BF01989313

Harvell CD, Mitchell CE, Jessica RW et al (2002) Climate warming and disease risks for terrestrial and marine biota. Science 296:2158–2162. https://doi.org/10.1126/science.1063699

Hawkins AJS, Bayne BL (1992) No title. In: Gosling E (ed) The mussel Mytilus: ecology, physiology, genetics and culture. Elsevier, Amsterdam, pp 171–222

Heisterkamp IM, Schramm A, Larsen LH et al (2013) Shell biofilm-associated nitrous oxide production in marine molluscs: processes, precursors and relative importance. Environ Microbiol 15:1943–1955. https://doi.org/10.1111/j.1462-2920.2012.02823.x

Hervio-Heath D, Colwell RR, Derrien A et al (2002) Occurrence of pathogenic vibrios in coastal areas of France. J Appl Microbiol 92:1123–1135. https://doi.org/10.1046/j.1365-2672.2002.01663.x

Hily C (1974) Développement larvaire de Mytilus edulis en présence d'antibiotiques. Rev Intern Océanogr Méd, 17 pp

Hine P (1999) The inter-relationships of bivalve haemocytes. Fish Shellfish Immunol 9:367–385. https://doi.org/10.1006/fsim.1998.0205

Høi L, Larsen JL, Dalsgaard I, Dalsgaard A (1998) Occurrence of Vibrio vulnificus biotypes in Danish marine environments. Appl Environ Microbiol 64:7–13

Holben WE, Williams P, Saarinen M et al (2002) Phylogenetic analysis of intestinal microflora indicates a novel Mycoplasma phylotype in farmed and wild salmon. Microb Ecol 44:175–185. https://doi.org/10.1007/s00248-002-1011-6

Holmström C, Kjelleberg S (1999) Marine Pseudoalteromonas species are associated with higher organisms and produce biologically active extracellular agents. FEMS Microbiol Ecol 30:285–293

Irianto A, Austin B (2002) Probiotics in aquaculture. J Fish Dis 25:633–642

Iwamoto M, Ayers T, Mahon BE, Swerdlow DL (2010) Epidemiology of seafood-associated infections in the United States. Clin Microbiol Rev 23:399–411

Jeanthon C, Prieur D, Cochard JC (1988) Bacteriological survey of antibiotic-treated sea waters in a Pecten maximus hatchery. Aquaculture 71:1–8. https://doi.org/10.1016/0044-8486(88)90267-0

Jeffries VE (1982) Three Vibrio strains pathogenic to larvae of Crassostrea gigas and Ostrea edulis. Aquaculture 29:201–226. https://doi.org/10.1016/0044-8486(82)90136-3

Jorquera MA, Silva FR, Riquelme CE (2001) Bacteria in the culture of the scallop Argopecten purpuratus (Lamarck, 1819). Aquac Int 9:285–303. https://doi.org/10.1023/A:1020449324456

Karim M, Zhao W, Rowley D et al (2013) Probiotic strains for shellfish aquaculture: protection of eastern oyster, Crassostrea virginica, larvae and juveniles against bacterial challenge. J Shellfish Res 32:401–408. https://doi.org/10.2983/035.032.0220

Kaspar CW, Tamplin ML (1993) Effects of temperature and salinity on the survival of Vibrio vulnificus in seawater and shellfish. Appl Environ Microbiol 59:2425–2429

Kaysner CA, Tamplin ML, Wekell MM et al (1989) Survival of Vibrio vulnificus in shellstock and shucked oysters (Crassostrea gigas and Crassostrea virginica) and effects of isolation medium on recovery. Appl Environ Microbiol 55:3072–3079

Kellogg CA, Lisle JT, Galkiewicz JP (2009) Culture-independent characterization of bacterial communities associated with the cold-water coral Lophelia pertusa in the northeastern Gulf of Mexico. Appl Environ Microbiol 75:2294–2303. https://doi.org/10.1128/AEM.02357-08

Kesarcodi-Watson A, Kaspar H, Lategan MJ, Gibson LF (2009) Challenge of New Zealand Greenshell™ mussel Perna canaliculus larvae using two Vibrio pathogens: a hatchery study. Dis Aquat Organ 86:15–20. https://doi.org/10.3354/dao02100

Kesarcodi-Watson A, Miner P, Nicolas JL, Robert R (2012) Protective effect of four potential probiotics against pathogen-challenge of the larvae of three bivalves: Pacific oyster (Crassostrea gigas), flat oyster (Ostrea edulis) and scallop (Pecten maximus). Aquaculture 344–349:29–34. https://doi.org/10.1016/j.aquaculture.2012.02.029

Kiorboe T, Mohlenberg F, Nohr O (1981) Effect of suspended bottom material on growth and energetics in Mytilus edulis. Mar Biol 61:283–288. https://doi.org/10.1007/BF00401567

Krause E, Wichels A, Giménez L et al (2012) Small changes in pH have direct effects on marine bacterial community composition: a microcosm approach. PLoS One. https://doi.org/10.1371/journal.pone.0047035

Kreeger DA, Langdon CJ, Newell RIE (1988) Utilization of refractory cellulosic carbon derived from Spartina alterniflora by the ribbed mussel Geukensia demissa. Mar Ecol Prog Ser 42:171–179

Krol RM, Hawkins WE, Overstreet RM (1991) Rickettsial and mollicute infections in hepatopancreatic cells of cultured Pacific white shrimp (Penaeus vannamei). J Invertebr Pathol 57:362–370. https://doi.org/10.1016/0022-2011(91)90140-L

Kueh CSW, Chan KY (1985) Bacteria in bivalve shellfish with special reference to the oyster. J Appl Bacteriol 59:41–47. https://doi.org/10.1111/j.1365-2672.1985.tb01773.x

Kwan TN, Bolch CJS (2015) Genetic diversity of culturable Vibrio in an Australian blue mussel Mytilus galloprovincialis hatchery. Dis Aquat Organ 116:37–46. https://doi.org/10.3354/dao02905

Labreuche Y, Soudant P, Gonçalves M et al (2006) Effects of extracellular products from the pathogenic Vibrio aestuarianus strain 01/32 on lethality and cellular immune responses of the oyster Crassostrea gigas. Dev Comp Immunol 30:367–379. https://doi.org/10.1016/j.dci.2005.05.003

Langdon CJ, Newell RIE (1990) Utilization of detritus and bacteria as food sources by two bivalve suspension-feeders, the oyster *Crassostrea virginica* and the mussel *Geukensia demissa*. Mar Ecol Prog Ser 58:299–310. https://doi.org/10.3354/meps058299

Langille M, Zaneveld J, Caporaso JG et al (2013) Predictive functional profiling of microbial communities using 16S rRNA marker gene sequences. Nat Biotechnol 31:814–821. https://doi.org/10.1038/nbt.2676

Leclerc M (1996) Humoral factors in marine invertebrates. In: Rinkevich B, Muller WEG (eds) Invertebrate immunology. Springer, Berlin, pp 1–10

Lee RW, Childress JJ (1995) Assimilation of inorganic nitrogen by seep mytilid Ia, an undescribed deep-sea mussel containing methanotrophic endosymbionts: fate of assimilated nitrogen and the relation between methane and nitrogen assimilation. Mar Ecol Prog Ser 123:137–148. https://doi.org/10.3354/meps123137

Lee RJ, Younger AD (2002) Developing microbiological risk assessment for shellfish purification. Int Biodeterior Biodegrad 50(3.4):177–183

Leippe M, Renwrantz L (1988) Release of cytotoxic and agglutinating molecules by Mytilus hemocytes. Dev Comp Immunol 12:297–308. https://doi.org/10.1016/0145-305X(88)90006-7

Lilly DM, Stillwell RH (1965) Probiotics: growth-promoting factors produced by microorganisms. Science 147(3659):747–748

Liltved H, Cripps SJ (1999) Removal of particle-associated bacteria by prefiltration and ultraviolet irradiation. Aquacult Res 30:445–450. https://doi.org/10.1046/j.1365-2109.1999.00349.x

Liltved H, Hektoen H, Efraimsen H (1995) Inactivation of bacterial and viral fish pathogens by ozonation or UV irradiation in water of different salinity. Aquacult Eng 14:107–122. https://doi.org/10.1016/0144-8609(94)P4430-J

Lim HJ, Kapareiko D, Schott EJ et al (2011) Isolation and evaluation of new probiotic bacteria for use in shellfish hatcheries: I. Isolation and screening for bioactivity. J Shellfish Res 30:609–615. https://doi.org/10.2983/035.030.0303

Lodeiros C, Bolinches J, Dopazo CP, Toranzo AE (1987) Bacillary necrosis in hatcheries of Ostrea edulis in Spain. Aquaculture 65:15–29. https://doi.org/10.1016/0044-8486(87)90267-5

Lodeiros C, Freites L, Fernandez E et al (1989) Efecto antibiótico de tres bacterias marinas en la supervivencia de larvas de la vieira Pecten ziczac infectadas con el germen Vibrio anguillarum. Boletín del Inst Ocean la Univ Oriente 28:165–169

Longeon A, Peduzzi J, Barthélemy M et al (2004) Purification and partial identification of novel antimicrobial protein from marine bacterium Pseudoalteromonas species strain X153. Mar Biotechnol 6:633–641. https://doi.org/10.1007/s10126-004-3009-1

Maas EW, Law CS, Hall JA et al (2013) Effect of ocean acidification on bacterial abundance, activity and diversity in the Ross Sea, Antarctica. Aquat Microb Ecol 70:1–15. https://doi.org/10.3354/ame01633

Manefield M, Turner SL (2002) Quorum sensing in context: out of molecular biology and into microbial ecology. Microbiology 148:3762–3764

Manefield M, Whiteley AS (2007) Acylated homoserine lactones in the environment: chameleons of bioactivity. Philos Trans R Soc Lond B Biol Sci 362:1235–1240. https://doi.org/10.1098/rstb.2007.2048

Mangwani N, Dash HR, Chauhan A, Das S (2012) Bacterial quorum sensing: functional features and potential applications in biotechnology. J Mol Microbiol Biotechnol 22:215–227

Martinez-Urtaza J, Bowers JC, Trinanes J, DePaola A (2010) Climate anomalies and the increasing risk of Vibrio parahaemolyticus and Vibrio vulnificus illnesses. Food Res Int 43:1780–1790. https://doi.org/10.1016/j.foodres.2010.04.001

Matranga V (1996) Molecular aspects of immune reactions in Echinodermata. In: Rinkevich B, Muller WEG (eds) Invertebrate immunology. Springer, Berlin, pp 235–247

Meron D, Atias E, Iasur Kruh L et al (2011) The impact of reduced pH on the microbial community of the coral Acropora eurystoma. ISME J 5:51–60. https://doi.org/10.1038/ismej.2010.102

Metchnikoff E (1908) The prolongation of life; optimistic studies. G.P. Putnam's Sons, New York

Miller MB, Bassler BL (2001) Quorum sensing in bacteria. Annu Rev Microbiol 55:165–199. https://doi.org/10.1146/annurev.micro.55.1.165

Mitta G, Vandenbulcke F, Roch P (2000) Original involvement of antimicrobial peptides in mussel innate immunity. FEBS Lett 486:185–190

Moreira R, Balseiro P, Forn-Cuní G et al (2018) Bivalve transcriptomics reveal pathogen sequences and a powerful immune response of the Mediterranean mussel (Mytilus galloprovincialis). Mar Biol 165:61. https://doi.org/10.1007/s00227-018-3308-0

Murchelano RA, Brown C, Bishop J (1975) Quantitative and qualitative studies of bacteria isolated from sea water used in the laboratory culture of the American oyster, Crassostrea virginica. J Fish Res Board Can 32:739–745. https://doi.org/10.1139/f75-096

Muroga K, Takahashi KG (2007) Overview on the studies of humoral defense factors in bivalve molluscs. Fish Pathol 42:1–17. https://doi.org/10.3147/jsfp.42.1

Nakamura A, Takahashi KG, Mori K (1999) Vibriostatic bacteria isolated from rearing seawater of oyster brood stock: potentiality as biocontrol agents for vibriosis in oyster larvae. Fish Pathol 34:139–144. https://doi.org/10.3147/jsfp.34.139

Oelschlaeger TA (2010) Mechanisms of probiotic actions—a review. Int J Med Microbiol 300:57–62

Olafsen JA, Mikkelsen HV, Glaever HM, Hansen GH (1993) Indigenous bacteria in hemolymph and tissues of marine bivalves at low temperatures. Appl Environ Microbiol 59:1848–1854

Paillard C, Le Roux F, Borrego JJ (2004) Bacterial disease in marine bivalves, a review of recent studies: trends and evolution. Aquat Living Resour 17:477–498. https://doi.org/10.1051/alr:2004054

Pezzati E, Canesi L, Damonte G et al (2015) Susceptibility of Vibrio aestuarianus 01/032 to the antibacterial activity of Mytilus haemolymph: identification of a serum opsonin involved in mannose-sensitive interactions. Environ Microbiol 17:4271–4279. https://doi.org/10.1111/1462-2920.12750

Pfister CA, Meyer F, Antonopoulos DA (2010) Metagenomic profiling of a microbial assemblage associated with the California mussel: a node in networks of carbon and nitrogen cycling. PLoS One. https://doi.org/10.1371/journal.pone.0010518

Pfister CA, Gilbert JA, Gibbons SM (2014) The role of macrobiota in structuring microbial communities along rocky shores. PeerJ 2:e631. https://doi.org/10.7717/peerj.631

Philipp EER, Kraemer L, Melzner F et al (2012) Massively parallel RNA sequencing identifies a complex immune gene repertoire in the lophotrochozoan Mytilus edulis. PLoS One 7:e33091. https://doi.org/10.1371/journal.pone.0033091

Pila EA, Sullivan JT, Wu XZ et al (2016) Haematopoiesis in molluscs: a review of haemocyte development and function in gastropods, cephalopods and bivalves. Dev Comp Immunol 58:119–128. https://doi.org/10.1016/j.dci.2015.11.010

Pino-Querido A, Álvarez-Castro JM, Vera M et al (2015) A molecular tool for parentage analysis in the Mediterranean mussel (*Mytilus galloprovincialis*). Aquacult Res 46:1721–1735. https://doi.org/10.1111/are.12329

Pipe RK (1990) Hydrolytic enzymes associated with the granular haemocytes of the marine mussel Mytilus edulis. Histochem J 22:595–603

Powell SM, Chapman CC, Bermudes M, Tamplin ML (2013) Dynamics of seawater bacterial communities in a shellfish hatchery. Microb Ecol 66:245–256. https://doi.org/10.1007/s00248-013-0183-6

Prado S, Montes J, Romalde JL, Barja J (2009) Marine activity of Phaeobacter strains against aquaculture pathogenic bacteria. Int Microbiol 12:107–114

Prado S, Romalde JL, Barja JL (2010) Review of probiotics for use in bivalve hatcheries. Vet Microbiol 145:187–197

Prieur D, Mevel G, Nicolas JL et al (1990) Interactions between bivalve molluscs and bacteria in the marine environment. Oceanogr Mar Biol Annu Rev 28:277–352

Prins TC, Smaal AC, Dame RF (1998) A review of the feedbacks between bivalve grazing and ecosystem processes. Aquat Ecol 31:349–359. https://doi.org/10.1023/a:1009924624259

Raszl SM, Froelich BA, Vieira CRW et al (2016) Vibrio parahaemolyticus and Vibrio vulnificus in South America: water, seafood and human infections. J Appl Microbiol 121:1201–1222

Renwrantz L (1983) Involvement of agglutinins (lectins) in invertebrate defense reactions: the immuno-biological importance of carbohydrate-specific binding molecules. Dev Comp Immunol 7:603–608. https://doi.org/10.1016/0145-305X(83)90074-5

Richards GP (1991) Shellfish depuration. In: Ward DR, Hackney CR (eds) Microbiology of marine food products. Van Nostrand Reinhold, New York, pp 395–428

Richardson LB, Burton DT, Stavola AM (1982) A comparison of ozone and chlorine toxicity to three life stages of the American oyster Crassostrea virginica. Mar Environ Res 6:99–113. https://doi.org/10.1016/0141-1136(82)90024-1

Riquelme C, Hayashida G, Araya R et al (1996) Isolation of a native bacterial strain from the Scallop Argopecten purpuratus with inhibitory effects against pathogenic vibrios. J Shellfish Res 15:369–374

Riquelme C, Araya R, Vergara N et al (1997) Potential probiotic strains in the culture of the Chilean scallop Argopecten purpuratus (Lamarck, 1819). Aquaculture 154:17–26. https://doi.org/10.1016/S0044-8486(97)00043-4

Riquelme CE, Jorquera MA, Rojas AI et al (2001) Addition of inhibitor-producing bacteria to mass cultures of Argopecten purpuratus larvae (Lamarck, 1819). Aquaculture 192:111–119. https://doi.org/10.1016/S0044-8486(00)00461-0

Robert R, Miner P, Nicolas JL (1996) Mortality control of scallop larvae in the hatchery. Aquacult Int 4:305–313. https://doi.org/10.1007/bf00120947

Roderick GE, Schneider KR (1994) Depuration and relaying of molluscan shellfish. In: Hackney CR, Pierson MD (eds) Environment indicators and shellfish safety. Chapman & Hall, New York

Rodríguez I, Alfonso A, Alonso E et al (2017) The association of bacterial C 9 -based TTX-like compounds with Prorocentrum minimum opens new uncertainties about shellfish seafood safety. Sci Rep 7:40880. https://doi.org/10.1038/srep40880

Rojas R, Miranda CD, Opazo R, Romero J (2015) Characterization and pathogenicity of Vibrio splendidus strains associated with massive mortalities of commercial hatchery-reared larvae of scallop Argopecten purpuratus (Lamarck, 1819). J Invertebr Pathol 124:61–69. https://doi.org/10.1016/j.jip.2014.10.009

Romalde JL, Diéguez AL, Lasa A, Balboa S (2014) New Vibrio species associated to molluscan microbiota: a review. Front Microbiol 4:413. https://doi.org/10.3389/fmicb.2013.00413

Romero A, Del Mar CM, Forn-Cuni G et al (2014) Occurrence, seasonality and infectivity of Vibrio strains in natural populations of mussels Mytilus galloprovincialis. Dis Aquat Organ 108:149–163. https://doi.org/10.3354/dao02701

Rost B, Zondervan I, Wolf-Gladrow D (2008) Sensitivity of phytoplankton to future changes in ocean carbonate chemistry: current knowledge, contradictions and research directions. Mar Ecol Prog Ser 373:227–237. https://doi.org/10.3354/meps07776

Rubiolo JA, Lozano-Leon A, Rodriguez-Souto R et al (2018) The impact of depuration on mussel hepatopancreas bacteriome composition and predicted metagenome. Antonie Van Leeuwenhoek 111(7):1117–1129. https://doi.org/10.1007/s10482-018-1015

Ruiz-Ponte C, Samain JJF, Sánchez JLJ, Nicolas JJL (1999) The benefit of a Roseobacter species on the survival of scallop larvae. Mar Biotechnol 1:52–59. https://doi.org/10.1007/PL00011751

Sainz-Hernández JC, Maeda-Martínez AN (2005) Sources of Vibrio bacteria in mollusc hatcheries and control methods: a case study. Aquacult Res 36:1611–1618. https://doi.org/10.1111/j.1365-2109.2005.01386.x

Skindersoe ME, Ettinger-Epstein P, Rasmussen TB et al (2008) Quorum sensing antagonism from marine organisms. Mar Biotechnol 10:56–63. https://doi.org/10.1007/s10126-007-9036-y

Sohn S, Lundgren KM, Tammi K et al (2016) Efficacy of probiotics in preventing vibriosis in the larviculture of different species of bivalve shellfish. J Shellfish Res 35:319–328. https://doi.org/10.2983/035.035.0206

Sugita H, Tanaami H, Kobashi T, Deguchi Y (1981) Bacterial flora of coastal bivalves. Bull Jpn Soc Sci Fish 47:655–661

Sugumar G, Nakai T, Hirata Y et al (1998) Vibrio splendidus biovar II as the causative agent of bacillary necrosis of Japanese oyster Crassostrea gigas larvae. Dis Aquat Organ 33:111–118. https://doi.org/10.3354/dao033111

Summerfelt ST (2003) Ozonation and UV irradiation—an introduction and examples of current applications. Aquacult Eng 28(1–2):21–36. https://doi.org/10.1016/S0144-8609(02)00069-9

Tanaka R, Ootsubo M, Sawabe T et al (2004) Biodiversity and in situ abundance of gut microflora of abalone (Haliotis discus hannai) determined by culture-independent techniques. Aquaculture 241:453–463. https://doi.org/10.1016/j.aquaculture.2004.08.032

Tannock GW (1997) Modification of the normal microbiota by diet, stress, antimicrobial agents, and probiotics. In: Mackie RI, With BA, Isaacson RE (eds) Gastrointestinal microbiology, vol. 2: Gastrointestinal microbes and host interactions, Chapman and Hall microbiology series. International Thomson Publishing, New York, pp 434–465

Travers M-A, Boettcher Miller K, Roque A, Friedman CS (2015) Bacterial diseases in marine bivalves. J Invertebr Pathol 131:11–31. https://doi.org/10.1016/j.jip.2015.07.010

Vasconcelos GJ, Lee JS (1972) Microbial flora of Pacific oysters (Crassostrea gigas) subjected to ultraviolet-irradiated seawater. Appl Microbiol 23:11–16

Venier P, Varotto L, Rosani U et al (2011) Insights into the innate immunity of the Mediterranean mussel Mytilus galloprovincialis. BMC Genomics. https://doi.org/10.1186/1471-2164-12-69

Vera M, Martínez P, Poisa-Beiro L et al (2011) Genomic organization, molecular diversification, and evolution of antimicrobial peptide myticin-c genes in the mussel (Mytilus galloprovincialis). PLoS One. https://doi.org/10.1371/journal.pone.0024041

Verschuere L, Rombaut G, Sorgeloos P, Verstraete W (2000) Probiotic bacteria as biological control agents in aquaculture. Microbiol Mol Biol Rev 64:655–671. https://doi.org/10.1128/MMBR.64.4.655-671.2000

Vezzulli L, Previati M, Pruzzo C et al (2010) Vibrio infections triggering mass mortality events in a warming Mediterranean Sea. Environ Microbiol 12:2007–2019. https://doi.org/10.1111/j.1462-2920.2010.02209.x

Vezzulli L, Brettar I, Pezzati E et al (2012) Long-term effects of ocean warming on the prokaryotic community: evidence from the vibrios. ISME J 6:21–30

Vezzulli L, Stagnaro L, Grande C et al (2018) Comparative 16SrDNA gene-based microbiota profiles of the Pacific oyster (Crassostrea gigas) and the Mediterranean mussel (Mytilus galloprovincialis) from a shellfish farm (Ligurian Sea, Italy). Microb Ecol 75(2):495–504. https://doi.org/10.1007/s00248-017-1051-6

Viollier PH, Shapiro L (2004) Spatial complexity of mechanisms controlling a bacterial cell cycle. Curr Opin Microbiol 7:572–578

Weber B, Hasic M, Chen C et al (2009) Type VI secretion modulates quorum sensing and stress response in Vibrio anguillarum. Environ Microbiol 11:3018–3028. https://doi.org/10.1111/j.1462-2920.2009.02005.x

Welsh DT, Castadelli G (2004) Bacterial nitrification activity directly associated with isolated benthic marine animals. Mar Biol 144:1029–1037. https://doi.org/10.1007/s00227-003-1252-z

Witt V, Wild C, Anthony KRN et al (2011) Effects of ocean acidification on microbial community composition of, and oxygen fluxes through, biofilms from the Great Barrier Reef. Environ Microbiol 13:2976–2989. https://doi.org/10.1111/j.1462-2920.2011.02571.x

Wootton JT, Pfister CA, Forester JD (2008) Dynamic patterns and ecological impacts of declining ocean pH in a high-resolution multi-year dataset. Proc Natl Acad Sci USA 105:18848–18853. https://doi.org/10.1073/pnas.0810079105

Wright RT, Coffin RB, Ersing CP, Pearson D (1982) Field and laboratory measurements of bivalve filtration of natural marine bacterioplankton. Limnol Oceanogr 27:91–98. https://doi.org/10.4319/lo.1982.27.1.0091

Feed Additives, Gut Microbiota, and Health in Finfish Aquaculture

Seyed Hossein Hoseinifar, Hien Van Doan, Maryam Dadar, Einar Ringø, and Ramasamy Harikrishnan

Abstract A complex and diverse microbial population inhibits the fish gut forming so-called gut microbiota. There were increasing research attempts regarding identification of the intestinal microbiota of fish. However, those studies were conducted using culture-based methods. To resolve the issues of those conventional methods for accurate taxonomic identification, molecular studies have been developed since 2000. The literature revealed important function of gut microbiota in mediating and stimulating gastrointestinal (GI) development, aiding digestive enzyme activity, maintaining mucosal tolerance, immunomodulation, and improving disease resistance. Besides, the production of various metabolites such as short-chain fatty acids (SCFAs) by fish gut microbiota was reported to affect digestive tract physiology and functions. Over the past two decades, a wide range of studies have been published on modulation of the microbial community of the finfish intestine using pro-, pre-, and synbiotics. This chapter provides an updated view on available literature regarding the effects of microbial feed additives on performance, immune response, as well as disease resistance of different finfish species. Also, we have highlighted the mode of

Seyed Hossein Hoseinifar and Ramasamy Harikrishnan contributed equally to the chapter book.

S. H. Hoseinifar (✉)
Department of Fisheries, Faculty of Fisheries and Environmental Sciences, Gorgan University of Agricultural Sciences and Natural Resources, Gorgan, Iran
e-mail: hoseinifar@gau.ac.ir

H. Van Doan
Department of Animal and Aquatic Sciences, Faculty of Agriculture, Chiang Mai University, Chiang Mai, Thailand

M. Dadar
Razi Vaccine and Serum Research Institute, Agricultural Research, Education and Extension Organization (AREEO), Karaj, Iran

E. Ringø
Norwegian College of Fishery Science, Faculty of Bioscience, Fisheries and Economics, UiT The Arctic University of Norway, Tromsø, Norway

R. Harikrishnan
Department of Zoology, Pachaiyappa's College for Men, Kanchipuram, Tamil Nadu, India

N. Derome (ed.), *Microbial Communities in Aquaculture Ecosystems*,
https://doi.org/10.1007/978-3-030-16190-3_6

actions as well as the gaps of knowledge regarding microbial feed additives and explored the topics that merit further investigations in the field.

1 An Introduction on the Importance of Gut Microbiota and Its Modulation

In this chapter, the microbiota is defined as the collective microbial community inhabiting a specific environment: the fish gut. The fish intestinal microbiota is different from human as the density and diversity may be similar in both the proximal and distal intestine. Even though fish represent the greatest diversity of all vertebrates, there is limited available information about composition of gut microbiota and its importance compared to terrestrial vertebrates.

In the 1930s, some researchers tried to perform studies on fish intestinal microbiota. Even though this and later information were available, in the 1970s, there was disagreement between scientists regarding the presence and the roles of indigenous gut microbiota in fish (Ringø et al. 2016). This has however changed, and the function of fish gut microbiota in mediating and stimulating gastrointestinal (GI) development, aiding digestive enzyme activity, maintaining mucosal tolerance, immunomodulation, and elevation of disease resistance has been confirmed (e.g., Hoseinifar et al. 2017a–d; Li et al. 2018). Thus, increasing our understanding about composition and structure of gut microbiota and the mode of actions behind their effects on fish growth performance and health status is of high importance. When discussing the crucial functions of the gut microbiota, it is relevant to consider the major role of indigenous bacteria in disease prevention, because the intestine is the infection route of several pathogens (e.g., Ringø et al. 2003, 2007).

Recently, Lescak and Milligan suggested a controversial statement about using teleost fish as a vertebrate model for understanding the interactions between gut microbiota and host, as traditional models relying on mammals (e.g., mouse, rat, dog) are usually limited by strains and sample sizes, thus compromising statistical power. In addition, mammal models focus on fecal samples (mostly composed of allochthonous strains) and not on the adherent (autochthonous) microbiota. Furthermore, different segments of the teleost GI tract must be investigated separately, as there may be different colonization patterns between beneficial bacteria and pathogens (Ringø et al. 2003). Furthermore, complex biotic and abiotic factors such as host genotype and the environment control host–microbe interactions.

From the 1930s and until the 1990s, there were several research attempts regarding characterization of intestinal microbiota of fish. The studies were based on culturing bacteria on medium and using physiological and biochemical characteristics to identify bacteria (Llewellyn et al. 2014). The main issues of those techniques were taking much time for identification as well as limited accuracy for identification of close strains. Besides, considering that above 95% of bacterial strains do not grow on culture medium, it was not possible to have a comprehensive picture of a whole bacterial community. To resolve this issue, molecular techniques were developed (from 2000) and used for identification of gut microbiota in different fish species (e.g., Ringø et al. 2006). However, in several studies, a high proportion of cultivable gut bacteria was identified as unknown strains by 16S rRNA gene sequencing. At present, culture-independent techniques are developed and commonly used for bacterial identification (e.g., Llewellyn et al. 2014, 2015; Ringø et al. 2016). The use of next-generation sequencing as novel technologies has broadened the current information regarding the taxonomic composition of indigenous microbiota in fish gut and revealed that a complex community of bacteria inhabits the intestine which was not believed before. For example, Ringø et al. stated in their review on salmonids, "*the gastrointestinal microflora of fish seems to be simpler than that of endotherms*," and it is a paradox that even today this statement is cited.

The production of various metabolites such as short-chain fatty acids (SCFAs) by fish gut microbiota has been reported. The main factors affecting production of metabolites are food intake and diet-mediated changes in the gut microbiota (e.g., Hoseinifar et al. 2017a–d; Nawaz et al. 2018). Acetate, propionate, and butyrate are the most abundant SCFAs in the gut which have beneficial effects on digestive tract physiology, gut health, and mucosal immune responses (Hoseinifar et al. 2016a–c, 2017a–d; Safari et al. 2016) as well as participation in various host-signaling mechanisms. Besides, they can serve as energy source for intestinal epithelium resulting in more growth and development of adsorptive surface which subsequently increases feed utilization efficiency (Hoseinifar et al. 2016a–c). In addition, during the last decade studies on endotherms revealed that SCFAs might have a key function for the prevention and treatment of the metabolic syndrome, bowel disorders, and certain types of cancers. For instance, extrathymic generation of T-reg cells has been reported to be caused by butyrate produced by intestinal microbiota during starch fermentation.

In order to evaluate the function of the gut microbiota in disease prevention, the researchers eliminate bacteria from the gut by using broad-spectrum antibiotics and germ-free animals. Since the publication of the first study on mice that revealed possible effects of the indigenous microbiota on the development of brain plasticity and subsequent physiological system response, several investigations have been performed on the gut–brain axis (e.g., Collins et al. 2012). Today, it is well known that a stable and healthy gut microbiota plays important roles in maintenance of normal status for digestive system and contributes to proper signaling along the gut–brain axis. Overall, an undisturbed gut microbiota will contribute to secure the healthy status of its host.

Over the past two decades, a wide range of studies have been published on modulation of the microbial community of the fish intestine (e.g., Ringø and Song 2016; Ringø et al. 2016). In addition, the effects of gut microbiota on carbohydrate and lipid metabolisms and metabolite profile are investigated. Although the fish intestinal microbiota is considerably understudied when compared with humans and endothermic animals, an impressive amount of knowledge on gut microbiota of fish and its importance has been published.

2 Gut Microbiota and Fish Immune Responses

2.1 Immune System

The fish immune system can be divided into innate (nonspecific) and adaptive (specific or memory), which are further subdivided into cellular or cell-mediated defense mechanisms and humoral defense mechanisms. These two parts of the fish immune system provide protection against a wide range of pathogens as well as foreign substances such as toxins and endogenous or exogenous components. The recognition of pathogen-associated molecular patterns (PAMPs) receptors results in activation of immune system. The main receptors are soluble forms such as LPS-binding protein, pentraxins, complement, and collectins.

2.1.1 Innate (Nonspecific) Immune System

The innate immune system (IIS) is the ability to eliminate foreign matter including microorganism and toxins based on the ability to differentiate self from non-self cells directly using different cellular and humoral components or responses (Magnadóttir 2006). This is a fundamental building block of immunology. Considering limited development of adaptive immune system (AIR), innate immune responses (IIR) are of high importance in fishes. The proliferation and maturation of antibodies in the poikilotherms is slow and the memory of lymphocytes is limited (Whyte 2007). Thus, the fish immune system functions in a generalized ability (IIR) and a more complex defense (AIR) to clear a large amount of foreign matter. The IIS is generally classified into three compartments, namely: (1) physiochemical or epithelial or mucosal barriers such as scales, epithelial surface on gills, skin, and gut, or gastrointestinal (GI) tract with secreted mucus (Magnadóttir 2010); (2) the humoral immune components such as cell secretions of complement compound (CC), C-reactive protein (CRP), interferon (IFN), lysozyme, transferrin, lectins, and antimicrobial peptides (AMPs); and (3) the cellular components such as nonspecific cytotoxic cells or natural killer cells (NK cells), monocytes/macrophages, thrombocytes, granulocytes, neutrophils, or lymphocytes (Jansson 2002; Magnadóttir 2010; Rodriguez-Tovar et al. 2011).

Physical barriers It has been reported that a variety of effective bioactive molecules such as lectins, pentraxins (PTX), lysozymes, complement proteins (CPs), AMPs, antiproteases, natural antibodies, and immunoglobulin M (IgM) exist in mucus secreted from skin mucus or gills of fish (Alexander and Ingram 1992; Rombout et al. 1993; Aranishi and Nakane 1997; Boshra et al. 2006; Saurabh and Sahoo 2008; Magnadóttir 2010; Whyte 2007). They can inhibit pathogens. Besides, the fish epidermis is capable of reacting to attacks through different mechanisms such as thickening and cellular hyperplasia, osmosis, or defending cells, such as lymphocytes, granulocytes, macrophages, eosinophilic granular cells, NK cells, C-type lectin receptor (CTLR) or nonspecific cytotoxic cells (NSCCs), or

T-cytotoxic cells (T-lymphocytes or $CD8^+$), all of which being important for prevention of pathogen entry (Hibiya 1994; Sveinbjornsson et al. 1996; Ellis 2001; Fischer et al. 2006).

Innate cellular components Phagocytic cells such as macrophages and granulocytes are maintained in the internal spaces to help with the inflammatory reaction and the neutralization of microorganisms and their toxins. The inflammation is classified as an innate defense mechanism (IDM) mediated by multifaceted interactions of cellular and humoral compounds, which cause more permeability of blood capillaries and subsequent ease in migration of defense cells. The granulocytes are morphology features as neutrophil, eosinophil, and basophil cells are among the first defense cells which migrate to the inflammation site and fight against pathogens (Magnadóttir 2006). Thrombocytes are one of the important defense cells of fish and are responsible for phagocytosis (Tavares-Dias et al. 1999). The neutrophils and macrophages are the main cells which are responsible for phagocytosis processes in fish (Secombes and Fletcher 1992) and eliminate pathogens through production of reactive oxygen species (ROS): so-called respiratory burst (RB) activity.

The innate cytotoxic cells can directly cause degradation of pathogens via pattern recognition and APCs, such as dendritic cells (DCs) (Magnadóttir 2006). Besides elimination of pathogens via phagocytosis, the macrophages are capable of acting as APCs for the production of antigens in the AIR. Furthermore, macrophages make use of cell-mediated exocytosis for superoxide radical production and subsequent elimination of pathogens through respiratory burst (RB) activity (Whyte 2007). The lymphocytes cells (B- and T-lymphocytes) are the main origin of immune responses, which are derived from bone marrow and thymus. The NK cells contain a C-type lectin receptor which is capable of stimulation and inhibition of cytokine production when it binds to MHC class I marked cells (Sato et al. 2003). The T-cytotoxic cells are known to identify virus-infected cells: the same function as human cluster of differentiation 8 ($CD8^+$) T-lymphocytes (Miller et al. 1998). The nonspecific cytotoxic cells (NSCCs), commonly known as NK cells or agranular small lymphocytes functions, can eliminate a range of spontaneously xenogeneic targets, such as fish parasites (Evans and Jaso-Friedmann 1992; Hasegawa et al. 1998).

Innate humoral components A key defense mechanism of humoral immune system (HIS) in fish is elimination of bacteria and their related toxins via the complement system. The humoral immune parameters have been classified according to their pattern recognition specificities or effectors function. The components eliminate or prevent growth and spread of pathogens through different ways. Also, there are other components which ease removal of pathogens via agglutinins (aggregate cells), precipitins (aggregate molecules), and opsonins (binding with pathogens).

The complement system includes a set of soluble and membrane proteins in inactive form that are activated by the sequential pathway of the innate immune through three complement pathways: (1) the classical pathway, binding antibody to pathogen's cell surface and production of a specific IgM; (2) the alternative pathway, direct activation by foreign microorganisms; and (3) the lectin pathway, the immune response produced by binding of a mannose-binding lectin (MBL) protein complex to mannans on the surface of the bacterial cell (Sakai 1992). These pathways are somewhat conserved in several vertebrates (Magor and Magor 2001).

The communication between the innate and adaptive immune systems (IIS and AIS) occurs via cytokines, which are signaling molecules. They are capable of regulating a variety of cells in an autocrine, paracrine, and endocrine fashion for immune effectors. Interleukins (ILs), IFN, tumor necrosis factors (TNFs), and transforming growth factors (TGFs) are among the main cytokine families (Secombes et al. 2009). The production of cytosines has been reported from several immune cells such as lymphocytes, granulocytes, and macrophages that trigger the immune response to limit the growth of both viruses and bacteria in fish (Savan and Sakai 2006). Moreover, various other lytic substances or hydrolase enzymes (e.g., cathepsin B, cathepsin L, chitinase, chitobiase, lysozyme, trypsin-like enzyme, serum amyloid P (SAP), lectins, antibodies, hemagglutinins, and inhibitory enzymes) play key roles in the elimination of pathogens.

2.1.2 Specific (Acquired) Immunity

The acquired specific immunity is triggered upon detection of an antigen which results from a previous reaction by immune system and the subsequent development of specific antibodies that will initiate reactions and culminate in the increase of circulation of specific antibodies, besides promoting immunological memory. The establishment of the specific immunity is quite slow, which requires specific receptor selection, cellular proliferation, and protein synthesis (Magnadóttir 2010).

Adaptive cellular components Antigen processed by antigen-presenting cells (APCs) will be presented to the T-lymphocytes which are capable of antigen characterization, strictly in the presence of a specific humoral component called major histocompatibility complex (MHC) molecules containing glycoprotein receptors. Upon establishment of antigen in the intracellular compartment, the cytotoxic T-lymphocytes perform the defense (Ellis 2001; Goldsby et al. 2002; Salinas et al. 2011). The two different lymphocytes cells (i.e., T and B) are involved in adaptive immunity. Besides, the fish-specific immunity also involves other lymphocyte cells named natural killers or T cytotoxic, which secrete cytokines as well as destroy the cells infected by viruses or injured cells (Tizard 2002; Raulet 2004). Unlike the innate immune system, the components of specific immune system (e.g., B- or and T-lymphocytes and Igs molecules or specific antibodies) are extremely specific to the antigen of the invading microbe. The B- or T-cells are responsible for the humoral and cell-mediated responses (Jansson 2002; Magnadóttir 2010).

Adaptive humoral components The key humoral immune component of the AIS is the Igs (antibodies) that are either expressed as B-lymphocyte receptors or secreted in plasma triggers for activation and proliferation of lymphocytes or in association with the MHC marker on APCs. The B-cells differentiate into long-lasting memory cells and plasma cells, which produce specific antibodies. Furthermore, T-cells are responsible for identification of pathogens, in association with the MHC marker on APCs (Jansson 2002; Magnadóttir 2010; Rodriguez-Tovar et al. 2011).

2.2 Gut Microbiota and Immunity

2.2.1 Gut Microbiota

The indigenous intestinal microbiota is central to the development and modulation of the mucosal innate and adaptive immune system and exerts important role in providing protection against pathogenic microbes by maintaining gut integrity and regulating intestinal barrier permeability (Fig. 1). Gut microbiota is continuously involved in facilitating digestion, activating metabolic functions, secreting vitamins, storing nutrients, and shaping intestinal architecture (Wang and Li 2015). In humans, gut microbiota is composed of various microbial populations, the most widespread being in the phyla of *Firmicutes* and *Bacteroidetes* representing about 80–90% of the whole gut microbiota (Lucas López et al. 2017). Gut microbiota populations are separated from intestinal epithelial cells by a physical-chemical barrier composed of mucus, mucin glycoproteins, and several antibacterial molecules, such as α-defensins, C-type lectins, lysozyme, phospholipase A2, and secretory IgA (Gallo and Hooper 2012). All gut microbial species in healthy conditions are in a mutualistic or commensal symbiotic state contributing to a perfect and constant homeostasis (Frosali et al. 2015). Gut microbiota with distinct features have a significant impact on the emergent immune system because it indicates a close interaction between gut microbiota and host defense mechanisms.

2.2.2 Gastrointestinal (GI) Tract

In general, the gastrointestinal (GI) tract or gut is continuously making contact with an overwhelming antigenic load such as indigenous bacteria and dietary antigens. Therefore, it is of high importance to discriminate commensal microbiota or dietary antigens from those antigens which need immune response (Wershil and Furuta 2008). GI tract is colonized by a number of microorganisms such as bacteria, archaea, virus, protozoan, and fungi. Besides beneficial bacteria in intestinal microbiota, there are pathogenic ones, which can cause diseases such as gut inflammation and invasiveness. The interaction between intestinal microbiota, epithelium, and mucosal immunity underlies a local or systemic homeostasis (Matamoros et al. 2013). Though, alteration in the balance between commensal and pathogenic bacteria results in disruption of homeostasis: so-called dysbiosis process (Littman and Pamer 2011), which subsequently causes local infection and inflammation (Maynard et al. 2012). The negative effect of stress on fish has been reported to cause immunosuppression, in addition to promoting dysbiosis in fish microbiome. Yet,

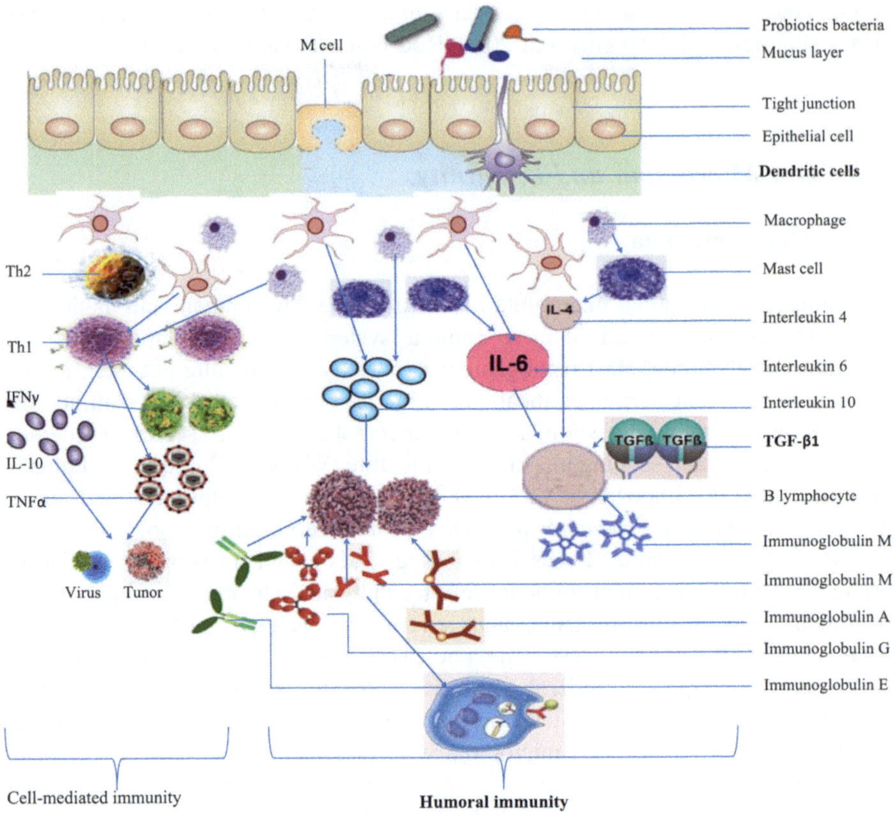

Fig. 1 Mechanism of action of probiotics for immunomodulation. The interaction of probiotic bacteria with epithelial cells (EC), M cells (MC), and dendritic cells (DC) results in the internalization of the bacteria or its components. This stimulates the release of IL-6 by EC and stimulates macrophages (MQ) and DC to induce the release of TNF-α and IFN-γ. At the same time, the mast cells (MAC) were also stimulated to release IL-4, which combined with IL-6 and TGF-β induced T-independent switch from IgM to IgA on the surface of B-lymphocytes (BL), enhancing the production of IgA. The IL-6 favors the clonal development of IgA B-lymphocytes which results in an increase in the production of IgM, IgG, and IgE antibodies. The Th1 cells produce pro-inflammatory IFNγ, TNFα, and IL-2 cytokines, which enhance or induce phagocytosis, macrophages, natural killer (NK) cells, and cytotoxic T-lymphocytes (CTL) to kill or inactivate viruses and tumors as well as eradicate infectious pathogens

the mode of action of stressors on alteration of gut microbiota and the impact of dysbiosis are not clear.

The GI tract contains a diverse and complex population of indigenous microbes, which produces acids, bile salts, and enzymes, creating antagonistic environment for pathogenic microorganisms (Li et al. 2018; Ringø et al. 2018). Generally, such characteristics are enough to provide protection against pathogenic microorganisms. If the conditions become favorable for the pathogenic microorganisms and their proliferation, cellular and humoral mechanisms would be activated as first defense lines to prevent spread of the infection. In this regard, the complement system has an important function and can be activated either directly by "foreign" agent surfaces or indirectly via other factors such as *C-reactive proteins* (CRP) and lectin. Besides complement system, there are other soluble factors such as antibacterial peptides (AMPs), proteases, and APPs like α2-macroglobulin, C3, lysozyme, lectins, PTX, and transferrin, which contribute to protection against disease.

The maturation of humoral immune mechanisms (i.e., circulating IgA- and IgM-secreting cells) affects colonization of indigenous microbiota (Grönlund et al. 2000). The integrity of the barrier function is maintained via luminal and mucosal parameters, achieved through restriction of pathogen colonization and avoiding entry of foreign agents into the mucosa (Sanderson and Walker 1993). It is well documented that microbial cell administered via feed in aquaculture have beneficial effects on fish performance, health, and resistance against disease, through modulation of microbial balance in the intestine toward potentially beneficial populations. (Merrifield et al. 2010; Hoseinifar et al. 2015a–c, 2016a–c, 2017a–d, 2018; Nawaz et al. 2018). In addition, they have other key functions such as competitive exclusion with harmful bacteria as well as production of bioactive metabolites, which interact with the immune system (Bloch et al. 2013).

2.2.3 Intestinal Barrier

The intestinal epithelium is covered by Peyer's patches (PPs), a specialized cell line known as M cells, which captures antigens and presents them to T- and B-cells. The antigen stimulation of lymphocytes causes proliferation of naive T- and B-cells, but it was activated in the digestive tract before migrating into the lymph vessels and mesenteric lymph nodes and finally entering into the bloodstream. At the same time, in the blood, T- and B-cells are transported back into the lymphatic structures of the digestive tract and the mucous membranes of other systems where they remain as effector cells. Intestinal microflora colonization by commensal bacteria stimulates the immune system, resulting in the formation of active PPs, proliferation of lymphocytes in the lamina propria, and increased production and circulation of secretory antibodies, mainly IgA and IgM. Nonpathogenic intestinal bacteria stimulate the formation of primarily natural antibodies, which are essential components of the innate immune mechanisms and constitute the first line of defense in the immune response (Cukrowska et al. 2001).

When looking at the interactions between intestinal microbiota and immunity, there are three distinct layers according to their structural aspects: (1) the layer composed of mucus (facing the intestinal lumen), including two sublayers—(a) the outer section is colonized by microbiota and (b) the inner section contains bactericidal AMPs and secretory IgA (SIgA) specific for commensal microorganisms (Maynard et al. 2012; Petersson et al. 2011; Johansson et al. 2011). (2) The second layer is mainly composed of a monolayer of intestinal epithelial cells (IECs) that are in contact with the lamina propria (LP) in their basolateral or apical surface with mucous layer. The IECs include a variety of cells such as mucin producing goblet cells, cholecystokinin and ghrelin producing enteroendocrine cells, AMPs producing Paneth cells, as well as M cells that capture antigens present in the immune system (Goto and Ivanov 2013; Collins et al. 2012). IECs have crucial functions such as separating the internal body organs from the outside environment through secretion of mucus and AMPs as well as the formation of tight junctions (Goto and Ivanov 2013). M cells are very important cells in IEC layers due to their direct interaction with the immune system, sampling antigens from lumen and carrying them in a unidirectional way to the APSc localized under the epithelium (Goto and Ivanov 2013). The enteroendocrine cells produce enteroendocrine peptide, namely glucagon-like peptide-2 (GLP-2) that provides protection in gut. GLP-2 is regulated by short-chain fatty acid production, induced by IEC proliferation. IEC proliferation upregulates tight junction protein genes in the intestine as well as modulates the immune system via expression of AMPs produced by Paneth cells (Cani et al. 2013). (3) The third layer is located beneath the IECs and includes gut-associated lymphoid tissues (GALT). The lamina propria (LP) contains mature isolated lymphoid follicles (ILFs). The sensing of microbe-associated molecular patterns (MAMPs) by PRRs on IECs results in the recruitment and activation of T- and B-lymphocytes in ILFs. Besides, PPs and ILFs contain several plasma cells which typically produce and release IgA (Kamada et al. 2013).

2.2.4 Immunomodulatory Roles of Microbiota

The colonization of intestinal mucosa with bacteria is regulated by both innate and adaptive immune pathways (Gallindo-Villegas et al. 2012; Bates et al. 2007). For instance, MyD88 signaling is of high importance regarding microbiota colonization control by the innate immune pathways (Gallindo-Villegas et al. 2012): MyD88 signaling is activated via MAMPs of intestinal microbiota strains. On the other side, some microbiota metabolites control the adaptive immune system: the sphingolipids produced by *Flectobacillus major* altered IgM and IgT levels of rainbow trout (*Oncorhynchus mykiss*) head kidney (HK) (Sepahi et al. 2016). Then, rainbow trout fed with probiotic containing diet exhibited a considerably increased serum total protein, serum albumin, IgM, and lysozyme (Kamgar et al. 2013). The probiotic strain *Enterococcus faecium* administered in Olive flounder (*Paralichthys olivaceus*) against *Lactococcus garvieae* elevated lysozyme activity, complement activity, and antiprotease activity (Kim et al. 2012). Also, supplementation of Nile tilapia diet with *Bacillus subtilis* and *Lactobacillus plantarum* probiotic mixture increased phagocytic activity, acid phosphatase activity, lysozyme activity, and total immunoglobulin activity. The probiotic strain *Lactobacillus acidophilus* administrated in African catfish (*Clarias gariepinus*) to fight against a pathogenic infection has improved serum total Igs concentration (Al-Dohail et al. 2011). In rainbow trout fed with probiotic diet, the cellular and humoral immune responses were enhanced. More specifically, the phagocytic activity of leukocytes and the alternative complement activity in serum were improved (Balcazar et al. 2007). Also, feeding rainbow trout with multi-strain probiotics (*L. rhamnosus*, *E. faecium*, and *B. subtilis*) modulated the expression immune-related genes as well as production of superoxide anions and leukocytes and alternate complement activity of serum (Panigrahi et al. 2007). In rainbow trout fed with probiotic diet, stimulation of the immune parameters as well as resistance against *Edwardsiella tarda* infection was observed (Newaj-Fyzul et al. 2007). Zorriehzahra et al. (2010) and Martins et al. (2008, 2009) also suggested that food supplemented with probiotics increases the total leukocyte count, lymphocytes, thrombocytes, and neutrophils in rainbow trout. Likewise, mixed probiotic feeding in rainbow trout for a period of 2 weeks stimulates the humoral and cellular immunity including lysozyme activity and results in an increase in the number of erythrocytes, macrophages, and lymphocytes and phagocytic activity, all of which representing early activation of the inflammatory response before antibody production (Irianto and Austin 2002).

Furthermore, the increase of anti-inflammatory cytokine secretion and subsequent reduction of inflammation were observed following probiotic treatment, therefore demonstrating the interaction of probiotic strain with epithelial cells. In a study with rainbow trout, Kim and Austin (2006) reported significant upregulation of IL-1β, IL-8, TNF-α, and TGF-β genes in the intestine, revealing induction of an anti-inflammatory effect (Kim and Austin 2006). Regarding the mode of action, it has been suggested that the receptors on immune cells (such as neutrophils, macrophages, and dendritic cells) recognize the β-glucans of probionts (Goodridge et al. 2009). Contact of β-glucans with TLRs results in induction of NF-κB and MAPK signaling (Gantner et al. 2003). Zymosan, a cell wall preparation of *Saccharomyces cerevisiae* including β-glucans, binds to TLR2 and TLR4 and via NF-κB pathway increases cytokine production (Gantner et al. 2003). Administration of probiotics in the gilthead seabream (*Sparus aurata*) notably increased immune parameters and upregulated immune-related genes such as Hep, IgM, TCR-ß, NCCRP-1, MHC-IIα, CSF-1R, C3, TNF-α, and IL-1ß (Reyes-Becerril et al. 2008). The interaction of β-glucans with specific receptors on macrophages and dendritic cells results in production of different cytokines, which in turn activate B- or T-lymphocytes and subsequently systemic immune response. It was suggested that Yeast β-glucans can modulate initially the innate immune system as long as the adaptive immune response is not sufficiently developed for protection against diseases (Bricknell and Dalmo 2005; Meena et al. 2013).

3 Modulation of Fish Gut Microbiota and Health Using Probiotics

It has been well documented that the intestinal microbiota of fish has important functions, providing healthiness, protection against diseases, as well as affecting maturation and development of the immune system. The normal microbiota has crucial effects on the innate immune system that is the key element for the fish disease resistance and is divided into cellular and humoral components as well as physical barriers.

Regarding the important effects of endogenous microbiota on the fish health status, numerous studies were done in this field. For instance, bacterial adherence has a critical role for colonization of pathogenic microorganisms (Hoseinifar et al. 2015a–c, 2017a–d). Therefore, the protective effects of three probionts derived from the fish intestine (*Lactobacillus plantarum* CLFP 238, *L. fermentum* CLFP 242, and *L. lactis* CLFP 101) were tested against the adhesion of different pathogens (*Vibrio anguillarum, Yersinia ruckeri, Aeromonas salmonicida*, and *A. hydrophila*) to intestinal mucosae (Balcázar et al. 2008). The outcomes of this study demonstrated that only *L. lactis* CLFP 101 reduced adhesion for the four fish pathogens. Furthermore, antibacterial effects of *L. lactis* CLFP 101 against the four pathogens were explored. In the case of *L. plantarum* CLFP 238, a protective effect against adhesion of *A. salmonicida* and *A. hydrophila* was noticed. In addition, the adhesion reduction of three out of four pathogens was observed for *L. fermentum* CLFP 242 administration. It is important to note that all LAB strains could survive in fish gut characterized by high bile concentrations and low pH, and all of them exerted an inhibition property of bacterial adherence. In another study, the effects of dietary *L. acidophilus* on performance, mucosal immunity, salinity stress resistance, as well as intestinal microbiota of black swordtail (*Xiphophorus helleri*) were investigated. The results showed the beneficial effects of *L. acidophilus*, indicating dietary use of probiotics with beneficial effects on the fish health and growth performance (Hoseinifar et al. 2015a–c). It has been revealed that lysozyme, protease, complements, lectins, and immunoglobulins are the main components of skin mucosal immunity, which protects fish against pathogens and plays an important role as the first defense line (Ángeles Esteban 2012). Similarly, in a study with dietary supplementation of *L. rhamnosus* or/and *L. lactis*, significantly increased growth performance and improved immune parameters were observed in red sea bream, *Pagrus major* (Dawood et al. 2016). Also, microbiological studies using culture-based method revealed a significant increase of total bacteria as well as lactic acid bacteria in intestinal microbiota (Dawood et al. 2016). Furthermore, Askarian et al. (2011) reported elevation of total protein level of skin mucus in the fry of Caspian white fish (*Rutilus frisii kutum*) following treatment with commercial probiotics (*L. casei*). It is thought that the increased skin-soluble protein content and skin mucus bactericidal activity can result from improved immune activities after dietary administration of probiotics. In addition, alkaline phosphatase (ALP) activity may play an important role in the innate immune system of fish, which may act as an antibacterial agent. On the other hand, the activity of ALP was reported to significantly increase when fish is treated with probiotic supplementation diet (Hoseinifar et al. 2015a–c). The elevated activity of alkaline phosphatase may be related to an improved mucosal immune response. In addition, the beneficial effects of dietary probiotic *L. curvatus* on growth, digestive enzyme, and intestinal microbiota of beluga (*Huso huso*) and Persian sturgeon (*Acipenser persicus*) fry have been reported (Askarian et al. 2011); the best results were obtained when fish fed 9×10^9 CFU g^{-1} level of probiotic bacteria. Similar positive effect was observed when 2×10^9 CFU g^{-1} *L. mesenteroides* was administered in Persian sturgeon diet, indicating the powerful ability of *Lactobacillus* spp. to colonize the digestive tract mucosa and induce health promotion (Askarian et al. 2011).

It has been demonstrated that treatment of parrot fish (*Oplegnathus fasciatus*) with *Bacillus subtilis* E20 could improve the growth performance and provide protection against *Vibrio alginolyticus* infection. It is important to note that the immune parameters of *O. fasciatus* were enhanced with elevation of probiotic levels, although the growth ratio was reduced. Also, feeding with probiotic supplemented diet at 10^{10} CFU kg^{-1} level notably increased resistance against *V. alginolyticus*. This has been attributed to elevation of immune parameters as observed with measurement of respiratory burst, phagocytic activity, and lysozyme activity, except for superoxide dismutase activity (Liu et al. 2018). Dietary formulation supplemented with 1.1×10^5 CFU g^{-1} *B. subtilis* could also significantly increase the weight gain (WG), final body weight (FBW), feed conversion ratio (FCR), protein efficiency ratio (PER), condition factor (K), specific growth rate (SGR) (%), energy retention (ER), and protein productive value (PPV, %) (Hassaan et al. 2018). Also, Hassaan et al. (2018) revealed that intestinal pH of *B. subtilis*-supplemented group was not affected and that total bacterial count in feces and gut decreased. The values of hemoglobin, hematocrit, white blood cells, red blood cells, albumin, globulin, and total protein were increased in *B. subtilis*-supplemented group. In another study, Soltan and El-Laithy (2008) showed that feeding with *B. subtilis*-supplemented diet reduced the AST and ALT activity in the Nile tilapia (*O. niloticus*), indicating a beneficial effect of *B. subtilis* in the regulation of the gut–liver axis as well as modulating intestinal microbiota composition (Soltan and El-Laithy 2008). Moreover, dietary *L. acidophilus* could regulate the protein profile of skin mucus, appetite, and immune genes expression in gold fish (*Carassius auratus gibelio*) by significantly increasing expression of TNF-1α and TNF-2α genes as well as downregulating ghrelin gene expression, while administration of various levels of *L. acidophilus* did not show any remarkable effect on growth performance (Hosseini et al. 2016).

4 Dietary Synbiotics Effects on Fish Gut Microbiota and Health Status

Combination of probiotics and prebiotics, termed as synbiotics, has revealed to be a promising means for disease biocontrol in farmed fish (Cerezuela et al. 2011; Hoseinifar et al. 2016a–c, 2017a–d, Van Doan et al. 2017). Overall, treatments of fish with synbiotics resulted in beneficial effects on immunological responses, survival, growth, gut microbiota, increased levels of intestinal absorption, and increased health status (González-Félix et al. 2018, Hassaan et al. 2018). It has been reported that *Bacillus* spp. and yeasts could act as an important probiotic with valuable properties that show effective roles in the growth and health of numerous cultured species (Li et al. 2009, Daniels et al. 2010, Cerezuela et al. 2011, Hosseini et al. 2016, Liu et al. 2018, Safari et al. 2018). Moreover, prebiotics such as galactooligosaccharides (GOS), isomaltooligosaccharides (IMO), mannanoligosaccharides (MOS), fructooligosaccharides (FOS), oligofructose, inulin, and several commercial prebiotics have been applied in several finfish (Gatlin III and Peredo 2012, González-Félix et al. 2018). Probiotics modulate digestion through the increase of both microbial enzymes and beneficial microbe activities (Hoseinifar et al. 2017a–d). They also improved intestinal microbiological features, indicating improved food absorption and digestion and gut microvilli morphology (Mohapatra et al. 2012).

The dietary supplementation of combined probiotics and prebiotics with commercial probiotic Aquablend® with a *Bacillus* strain at concentration of 11×10^9 CFU kg^{-1} and the yeast-based prebiotic GroBiotic®-A was tested in totoaba (*Totoaba macdonaldi*). Supplementation was either probiotic only or combined with the prebiotic. In both cases, there were no remarkable differences for growth and survival indices evaluated at the end of the experimental trials (González-Félix et al. 2018). In another study, the effects of combined or singular administration of *L. casei* as a probiotic and *Agaricus bisporus* extracts as a prebiotic in zebrafish (*Danio rerio*) were tested. First, lysozyme activity of the plasma in the fish fed with the prebiotic was lower when compared with that in control group. Furthermore, the fold length in the proximal area of gastrointestinal tract of probiotic fed fish was remarkably larger than control treatment. However, remarkable changes were found in the overall transient microbial diversity and community by V3-V4 16S rRNA gene massive sequencing, indicating various bacterial clustering profiles between the additives (probiotic, prebiotic, and combined probiotic and prebiotic) and the basal diet. In addition, the effective role of combined or singular administration of *L. casei* and *Agaricus bisporus* on skin mucus immune parameters as well as the expression of selected genes related to growth, appetite, mucosal immunity, and antioxidant enzymes revealed that oral administration of *A. bisporus* and *L. casei* significantly elevated growth-associated genes (*igf1* and *gh*), the mucosal immune-related genes (*lyz*, *tnf-alpha*, and *il1b*), as well as antioxidant-related genes (sod, cat) expression in zebrafish fed with combined diet supplemented with *A. bisporus* and *L. casei*. Moreover, skin mucus nonspecific immune factors in combined diet supplemented with *A. bisporus* and *L. casei* were significantly higher than those in *A. bisporus*, *L. casei*, or control groups, indicating the promising effects of combined administration of *A. bisporus* and *L. casei* as beneficial feed additive in fish aquaculture (Safari et al. 2018).

Hassaan et al. (2018) supplemented Nile tilapia (*Oreochromis niloticus*) diet with malic acid and *Bacillus subtilis*. The results revealed that the survival and growth value was higher in dietary treatments with malic acid and *Bacillus subtilis* in comparing with the control fed fish. The highest values of SGR (%), WG, FBW, PPV, PER, and ER were reported in dietary groups with 10 g malic acid kg^{-1} and 1.1×10^5 CFU g^{-1} *B. subtilis* and 5 g malic acid kg^{-1} and 1.1×10^5 CFU g^{-1} *B. subtilis*. In addition, the highest values of white blood cells, red blood cells, hemoglobin, hematocrit, total protein, albumin, and globulin were found in the dietary treatments with malic acid and *Bacillus subtilis* (Hassaan et al. 2018).

Regarding the mode of action, it has been suggested that lowering gastric pH lowers intestinal pH and, in turn, enhances the nutrient utilization and causes activation of pepsin and could increase mineral solubilization and its absorption (Hassaan et al. 2018).

In another study, the best synbiotic combination was determined between *Pediococcus acidilactici* and a range of prebiotics under in vitro conditions and based on bacterial growth and production of short-chain fatty acid (Hoseinifar et al. 2017a–d). Then, under in vivo conditions the possible effects of synbiotic were studied on growth performance, intestinal microbiota, and physiological response, of rainbow trout (*Oncorhynchus mykiss*) fingerlings. The outcome of the study revealed an increase of performance immune response and disease resistance in fish fed *P. acidilactici* and galactooligosaccharide (GOS) (Hoseinifar et al. 2015a–c). Also, regulation of disease resistance against *Streptococcus iniae*, skin mucosal activities (bactericidal activity against *Escherichia coli*, *Serratia marcescens*, *Streptococcus iniae*, *Streptococcus faecium*, and *Staphylococcus aureus* as well as the content of mucus protein), and innate immune response (alternative complement, respiratory burst activities and lysozyme) in rainbow trout fed with the synbiotic *P. acidilactici*/GOS were reported (Hoseinifar et al. 2015a–c). Similar to Hoseinifar and his colleagues' findings, the capacity of a symbiotic *Bacillus clausii*/MOS/FOS to improve immune activities was observed for Japanese flounder (*Paralichthys olivaceus*) comparatively to control groups fed with singularly *Bacillus clausii*, MOS, and FOS (Ye et al. 2011). Similarly, combined administration of *B. subtilis* (1.0 g kg^{-1}) and chitosan (6.0 g kg^{-1}) in cobia (*Rachycentron canadum*) (Geng et al. 2011), *Weissella cibaria*/inulin in hybrid surubim (*Pseudoplatystoma* sp.) (Mouriño et al. 2012), and *B. subtilis* (1.35 × 107 CFU g^{-1})/FOS in juvenile large yellow croaker (*Larimichthys crocea*) (Ai et al. 2011) showed beneficial effects on performance, diet utilization, innate immune parameters, and disease resistance.

It is important to note that the oral administration of inulin (0.5%) with *W. cibaria* in hybrid surubins (*Pseudoplatystoma corruscans* × *P. reticulatum*) decreased the presence of pathogenic bacteria and improved the intestinal microbiota that are related with their immune defense system (Mouriño et al. 2012). Although the accurate mechanisms of dietary synbiotics on fish health status need more detailed investigations, recent outcomes have speculated that they can affect immune parameters via production of SCFA following microbial fermentation. SCFAs could change the innate immune activities by binding to GPR43, G protein coupled receptor, which exists on immune cells (Maslowski and Mackay 2010). On the other hand, feeding with *B. subtilis*/chitosan significantly increased the serum alternative complement pathway (ACP) activities that are important in fish nonspecific immune responses (Geng et al. 2011).

Rodriguez-Estrada and his colleagues reported that dietary administration of MOS with *Enterococcus faecalis* in rainbow trout could increase growth ratio and stimulate the immune activities, although in *Japanese flounder* the same synbiotic formulation did not exert a clear synergistic effect (Rodriguez-Estrada et al. 2009). Similarly, no synergistic effects were reported between FOS and *B. subtilis* against *V. harveyi* in the yellow croaker (Hoseinifar et al. 2015a–c). The apparent contradictory outcomes could be associated with the possible administration of not suitable prebiotics as substrate for probionts, which result in no or limited fermentation and subsequent accumulation of prebiotics (Hoseinifar et al. 2015a–c).

Therefore, the most promising aspect of synbiotics studies in fish aquaculture is the necessity of evaluating prebiotic and probiotic safety by gaining knowledge on their effects on intestinal microbiota activity. In addition, a better public acceptance for probiotic and prebiotic rationale, more accurate guidelines regarding safety statements, and preparation could be of interest to accelerate the development of a more sustainable commercial aquaculture.

5 Research Gaps and Future Perspectives

Looking at available literature regarding prebiotics revealed contradictory results and in some cases negative effects. This has been attributed to the interspecific fish host differences regarding the composition of intestinal microbial communities, which was reported to be the main factor affecting fermentability and functionality of prebiotics. Therefore, before selection and administration of a prebiotic mixture, a thorough study on intestinal microbiota and fermentation ability should be performed. On the other hand, the degree of polymerization (DP) of prebiotics considerably affects fermentation by microbiota. For example, previous studies on the same species (i.e., Beluga, *Huso huso*) revealed that prebiotics with different DP (inulin and oligofructose) had different effects (Hoseinifar et al. 2011). Therefore, fermentation of prebiotics with various DPs by intestinal microbiota using in vitro and ex vivo studies should be considered (Hoseinifar et al. 2015a–c). Such studies would give helpful information to researchers to select proper prebiotics and optimum inclusion levels for cultured finfish.

As mentioned it has been well documented that gut microbiota of fish has crucial functions. Considering the importance of gut microbiota in digestive physiology and function as well as providing protection against disease, it is of high importance to consider the balance of microbial community and alteration toward beneficial bacteria. The chemicals such as antibiotics alter the microbial community and change the condition toward adhesion and colonization of harmful bacteria. Administration of probiotics in such conditions can be considered to help recovering gut microbiota functional homeostasis.

There is increasing literature regarding the cross talk between gut microbiota and mucosal immunity. The mucosal immunity can play a role as the first defense line against diseases. Therefore, strengthening of gut microbiota by using functional feed additives such as pro-, pre-, or synbiotics can benefit immunity. Also, although there are some hypotheses regarding their mode of action, more studies are still to be performed to confirm them. Metagenomics and transcriptomic studies can help to increase existing knowledge about the functionality of fish gut microbiota and their partnership with the fish immune system. On the other hand, the review of literature revealed more beneficial results when synbiotics are administered. However, there are very limited studies regarding determination of beneficial substrate for each probiotic and introduction of optimum synbiotic mixture. This can be considered as an obvious area of future research.

References

Ai Q, Xu H, Mai K, Xu W, Wang J, Zhang W (2011) Effects of dietary supplementation of Bacillus subtilis and fructooligosaccharide on growth performance, survival, non-specific immune response and disease resistance of juvenile large yellow croaker, *Larimichthys crocea*. Aquaculture 317:155–161

Al-Dohail MA, Hashim R, Aliyu-Paiko M (2011) Evaluating the use of 507 *Lactobacillus acidophilus* as a biocontrol agent against common pathogenic 508 bacteria and the effects of the haematology parameters and histopathology in 509 African catfish *Clarias gariepinus* juveniles. Aquacult Res 42:196–209

Alexander JB, Ingram GA (1992) Noncellular nonspecific defense mechanisms of fish. Annu Rev Fish Dis 2:249–279

Ángeles Esteban M (2012) An overview of the immunological defenses in fish skin. ISRN Immunol. https://doi.org/10.5402/2012/853470

Aranishi F, Nakane M (1997) Epidermal proteases of the Japanese eel. Fish Physiol Biochem 16:471–478

Askarian F, Kousha A, Salma W, Ringø E (2011) The effect of lactic acid bacteria administration on growth, digestive enzyme activity and gut microbiota in Persian sturgeon (*Acipenser persicus*) and beluga (*Huso huso*) fry. Aquacult Nutr 17:488–497

Balcazar JL, de Blas I, Ruiz-Zarzuela I, Vendrell D, Calvo AC, Marquez I, O G, Muzquiz JL (2007) Changes in intestinal microbiota and humoral immune response following probiotic administration in brown trout (*Salmo trutta*). Br J Nutr 97:522–527

Balcázar JL, Vendrell D, de Blas I, Ruiz-Zarzuela I, Muzquiz JL, Girones O (2008) Characterization of probiotic properties of lactic acid bacteria isolated from intestinal microbiota of fish. Aquaculture 278:188–191

Bates JM, Akerlund J, E M, Guillemin K (2007) Intestinal alkaline phosphatase detoxifies lipopolysaccharide and prevents inflammation in zebrafish in response to the gut microbiota. Cell Host Microbe 2:371–382

Bloch O, Crane CA, Kaur R, Safaee M, Rutkowski MJ, Parsa AT (2013) Gliomas promote immunosuppression through induction of B7-H1 expression in tumor-associated macrophages. Clin Cancer Res 19:3165–3175

Boshra H, Li J, Sunyer JO (2006) Recent advances on the complement system of teleost fish. Fish Shellfish Immunol 20:239–262

Bricknell I, Dalmo RA (2005) The use of immunostimulants in fish larval aquaculture. Fish Shellfish Immunol 19:457–472

Cani PD, Everard A, Duparc T (2013) Gut microbiota, enteroendocrine functions and metabolism. Curr Opin Pharmacol 13:935–940

Cerezuela R, Meseguer J, Esteban M (2011) Current knowledge in synbiotic use for fish aquaculture: a review. J Aquacult Res Dev 1:1–7

Collins SM, Surette M, Bercik P (2012) The interplay between the intestinal microbiota and the brain. Nat Rev Microbiol 10:735–742

Cukrowska B, Kozakova H, Rehakova Z, Sinkora J, Tlaskalova-Hogenova H (2001) Specific antibody and immunoglobulin responses after intestinal colonization of germ-free piglets with non-pathogenic *Escherichia coli* O86. Immunobiology 204:425–433

Daniels CL, Merrifield DL, Boothroyd DP, Davies SJ, Factor JR, Arnold KE (2010) Effect of dietary Bacillus spp. and mannan oligosaccharides (MOS) on European lobster (*Homarus gammarus* L.) larvae growth performance, gut morphology and gut microbiota. Aquaculture 304:49–57

Dawood MA, Koshio S, Ishikawa M, Yokoyama S, El Basuini MF, Hossain MS, Nhu TH, Dossou S, Moss AS (2016) Effects of dietary supplementation of *Lactobacillus rhamnosus* or/and *Lactococcus lactis* on the growth, gut microbiota and immune responses of red sea bream, Pagrus major. Fish Shellfish Immunol 49:275–285

Ellis AE (2001) Innate host defence mechanism of fish against viruses and bacteria. Dev Comp Immunol 25:827–839

Evans D, Jaso-Friedmann L (1992) Nonspecific cytotoxic cells as effectors of immunity in fish. Annu Rev Fish Dis 2:109–121

Fischer U, Utke K, Somamoto T, Kollner B, Ototake M, Nakanishi T (2006) Cytotoxic activities of fish leucocytes. Fish Shellfish Immunol 20:209–226

Frosali S, Pagliari D, Gambassi G, Landolfi R, Pandolfi F, Cianci R (2015) How the intricate interaction among toll like receptors, microbiota, and intestinal immunity can influence gastro-intestinal pathology. J Immunol Res Article ID 489821:1–12

Gallindo-Villegas J, Garcia-Moreno D, Oliveira S, Meseguer J, Mulero V (2012) Regulation of immunity and disease resistance by commensal microbes and chromatin modifications during zebrafish development. Proc Natl Acad Sci USA 109:E2605–E2614

Gallo RL, Hooper LV (2012) Epithelial antimicrobial defence of the skin and intestine. Nat Rev Immunol 12:503–516

Gantner BN, Simmons RM, Canavera SJ, Akira S, Underhill DM (2003) Collaborative induction of inflammatory responses by dectin-1 and toll-like receptor 2. J Exp Med 197:1107–1117

Gatlin III D, Peredo AM (2012) Prebiotics and probiotics: definitions and applications. Prebiotics and probiotics: definitions and applications

Geng X, Dong X-H, Tan B-P, Yang Q-H, Chi S-Y, Liu H-Y, Liu X-Q (2011) Effects of dietary chitosan and *Bacillus subtilis* on the growth performance, non-specific immunity and disease resistance of cobia, *Rachycentron canadum*. Fish Shellfish Immunol 31:400–406

Goldsby RA, Kindt TJ, BA O, Kuby J (2002) Immunology, 5th edn. WH Freeman, New York. 554 p

González-Félix ML, Gatlin DM III, Urquidez-Bejarano P, de la Reé-Rodríguez C, Duarte-Rodríguez L, Sánchez F, Casas-Reyes A, Yamamoto FY, Ochoa-Leyva A, Perez-Velazquez M (2018) Effects of commercial dietary prebiotic and probiotic supplements on growth, innate immune responses, and intestinal microbiota and histology of Totoaba macdonaldi. Aquaculture 491:239–251

Goodridge HS, Wolf AJ, Underhill DM (2009) β-glucan recognition by the innate immune system. Immunol Rev 230:38–50

Goto Y, Ivanov II (2013) Intestinal epithelial cells as mediators of the commensal-host immune crosstalk. Immunol Cell Biol 91:204–214

Grönlund MM, Arvilommi H, Kero P, Lehtonen OP, Isolauri E (2000) Importance of intestinal colonization in the maturation of humoral immunity in early infancy: a prospective follow up study of healthy infants aged 0-6 months. Arch Dis Child 83:F186–F192

Hasegawa S, Nakayasu C, Yoshitomi T, Nakanishi T, Okamoto N (1998) Specific cell-mediated cytotoxicity against an allogeneic target cell line in isogeneic ginbuna crucian carp. Fish Shellfish Immunol 8:303–313

Hassaan M, Soltan M, Jarmołowicz S, Abdo H (2018) Combined effects of dietary malic acid and Bacillus subtilis on growth, gut microbiota and blood parameters of Nile tilapia (*Oreochromis niloticus*). Aquacult Nutr 24:83–93

Hibiya T (ed) (1994) An atlas of fish histology. Normal and pathological features. Gustav Fischer, Stuttgart, pp 5–125

Hoseinifar SH, Mirvaghefi A, Mojazi Amiri B, Rostami HK, Merrifield DL (2011) The effects of oligofructose on growth performance, survival and autochthonous intestinal microbiota of beluga (*Huso huso*) juveniles. Aquacult Nutr 17:498–504

Hoseinifar SH, Esteban MÁ, Cuesta A, Sun Y-Z (2015a) Prebiotics and fish immune response: A review of current knowledge and future perspectives. Rev Fish Sci Aquacult 23:315–328

Hoseinifar SH, Mirvaghefi A, Amoozegar MA, Sharifian M, Esteban MÁ (2015b) Modulation of innate immune response, mucosal parameters and disease resistance in rainbow trout (*Oncorhynchus mykiss*) upon synbiotic feeding. Fish Shellfish Immunol 45:27–32

Hoseinifar SH, Roosta Z, Hajimoradloo A, Vakili F (2015c) The effects of *Lactobacillus acidophilus* as feed supplement on skin mucosal immune parameters, intestinal microbiota, stress resistance and growth performance of black swordtail (*Xiphophorus helleri*). Fish Shellfish Immunol 42:533–538

Hoseinifar SH, Khalili M, Sun Y-Z (2016a) Intestinal histomorphology, autochthonous microbiota and growth performance of the oscar (*Astronotus ocellatus* Agassiz, 1831) following dietary administration of xylooligosaccharide. J Appl Ichthyol 32:1137–1141

Hoseinifar SH, Ringø E, Shenavar Masouleh A, Esteban MÁ (2016b) Probiotic, prebiotic and synbiotic supplements in sturgeon aquaculture: a review. Rev Aquacult 8:89–102

Hoseinifar SH, Zoheiri F, Caipang CM (2016c) Dietary sodium propionate improved performance, mucosal and humoral immune responses in Caspian white fish (*Rutilus frisii kutum*) fry. Fish Shellfish Immunol 55:523–528

Hoseinifar S, Mirvaghefi A, Amoozegar M, Merrifield D, Ringø E (2017a) In vitro selection of a synbiotic and in vivo evaluation on intestinal microbiota, performance and physiological response of rainbow trout (*Oncorhynchus mykiss*) fingerlings. Aquacult Nutr 23:111–118

Hoseinifar SH, Dadar M, Ringø E (2017b) Modulation of nutrient digestibility and digestive enzyme activities in aquatic animals: the functional feed additives scenario. Aquacult Res 48:3987–4000

Hoseinifar SH, Safari R, Dadar M (2017c) Dietary sodium propionate affects mucosal immune parameters, growth and appetite related genes expression: Insights from zebrafish model. Gen Comp Endocrinol 243:78–83

Hoseinifar SH, Sun Y, Zhou Z (2017d) Prebiotics and synbiotics. In: Austin B, Newaj-Fyzul A (eds) Diagnosis and control of diseases of fish and shellfish. Wiley, Oxford

Hoseinifar SH, Sun Y, Wang A, Zhou Z (2018) Probiotics as means of diseases control in aquaculture, a review of current knowledge and future perspectives. Front Microbiol 9:2429

Hosseini M, Miandare HK, Hoseinifar SH, Yarahmadi P (2016) Dietary *Lactobacillus acidophilus* modulated skin mucus protein profile, immune and appetite genes expression in gold fish (*Carassius auratus gibelio*). Fish Shellfish Immunol 59:149–154

Irianto A, Austin B (2002) Probiotics in aquaculture. J Fish Dis 25:633–642

Jansson E (2002) Bacterial kidney disease in salmonid fish: Development of methods to assess immune functions in salmonid fish during infection by *Renibacterium salmoninarum*. PhD thesis, 1–52, Department of Pathology and Department of Fish, National Veterinary Institute, Swedish University of Agricultural Science, Uppsala, Sweden

Johansson M, Holmen Larsson J, Hansson G (2011) The two mucus layers of colon are organized by the MUC2 mucin, whereas the outer layer is a legislator of host-microbial interactions. Proc Natl Acad Sci USA 108:4659–4665

Kamada N, Seo SU, Chen GY, Núñez G (2013) Role of the gut microbiota in immunity and inflammatory disease. Nat Rev Immunol 13:321–335

Kamgar M, Pougholam R, Ghiasi M, Ghane M (2013) Studies on *Bacillus subtilis*, as potential probiotics on the biochemical parameters of rainbow trout, *Oncorhynchus mykiss* (Walbaum) to challenge infections. Adv Stud Biol 5:37–50

Kim DH, Austin B (2006) Innate immune responses in rainbow trout (*Oncorhynchus mykiss*, Walbaum) induced by probiotics. Fish Shellfish Immunol 21:513–524

Kim YR, Kim EY, Choi SY, Hossain MT, Oh R, Heo WS, Lee JM, Cho YC, Kong IS (2012) Effect of a probiotic strain, *Enterococcus faecium*, on the immune responses of Olive flounder (*Paralichthys olivaceus*). J Microbiol Biotechnol 22:526–529

Li J, Tan B, Mai K (2009) Dietary probiotic Bacillus OJ and isomaltooligosaccharides influence the intestine microbial populations, immune responses and resistance to white spot syndrome virus in shrimp (*Litopenaeus vannamei*). Aquaculture 291:35–40

Li X, Ringø E, Hoseinifar SH, Lauzon H, Birkbeck H, Yang D (2018) Adherence and colonisation of microorganisms in the fish gastrointestinal tract. Rev Aquacult. https://doi.org/10.1111/raq.12248

Littman DR, Pamer EG (2011) Role of the commensal microbiota in normal and pathogenic host immune responses. Cell Host Microbe 10:311–323

Liu C-H, Wu K, Chu T-W, Wu T-M (2018) Dietary supplementation of probiotic, *Bacillus subtilis* E20, enhances the growth performance and disease resistance against *Vibrio alginolyticus* in parrot fish (*Oplegnathus fasciatus*). Aquacult Int 26:63–74

Llewellyn MS, Boutin S, Hoseinifar SH, Derome N (2014) Teleost microbiomes: the state of the art in their characterization, manipulation and importance in aquaculture and fisheries. Front Microbiol 5:207

Llewellyn MS, McGinnity P, Dionne M, Letourneau J, Thonier F, Carvalho GR, Creer S, Derome N (2015) The biogeography of the Atlantic salmon (Salmo salar) gut microbiome. ISME J 10:1280–1284

Lucas López R, Grande Burgos MJ, Gálvez A, Pérez Pulido R (2017) The human gastrointestinal tract and oral microbiota in inflammatory bowel disease: a state of the science review. APMIS 125:3–10

Magnadóttir B (2006) Innate immunity of fish (overview). Fish Shellfish Immunol 20:137–151

Magnadóttir B (2010) Immunological control of fish diseases. J Mar Biotechnol 12:361–379

Magor BG, Magor KE (2001) Evolution of effectors and receptors of innate immunity. Dev Comp Immunol 25:651–682

Martins ML, Mouriño JL, Amaral GV, Vieira FN, Dotta G, Jatobá AM, Pedrotti FS, Jerônimo GT, Buglione-Neto CC, Pereira G Jr (2008) Haematological changes in Nile tilapia experimentally infected with *Enterococcus* sp. Braz J Biol 68:631–637

Martins FS, Veloso LC, Arantes RM, Nicoli JR (2009) Effects of yeast probiotic formulation on viability, revival and protection against infection with *Salmonella enterica* ssp. enterica serovar Typhimurium in mice. Lett Appl Microbiol 49:738–744

Maslowski KM, Mackay CR (2010) Diet, gut microbiota and immune responses. Nat Immunol 12:5

Matamoros S, Gras-Leguen C, Le Vacon F, Potel G, de La Cochetiere MF (2013) Development of intestinal microbiota in infants and its impact on health. Trends Microbiol 21:167–173

Maynard CL, Elson CO, Hatton RD, Weaver CT (2012) Reciprocal interactions of the intestinal microbiota and immune system. Nature 489:231–241

Meena DK, Das P, Kumar S, Mandal SC, Prusty AK, Singh SK, Akhtar MS, Behera BK, Kumar K, AK& P, Mukherjee SC (2013) Beta-glucan: an ideal immunostimulant in aquaculture (a review). Fish Physiol Biochem 39:431–457

Merrifield DI, Dimitroglou A, Foey A, Davies SJ, Baker RTM, Bøgwald J, Castex M, Ringø E (2010) The current status and future focus of probiotic and prebiotic applications for salmonids. Aquaculture 302:1–18

Miller N, Wilson M, Bengten E, Stuge T, Warr G, Clem W (1998) Functional and molecular characterization of teleost leukocytes. Immunol Rev 166:187–197

Mohapatra S, Chakraborty T, Prusty A, Das P, Paniprasad K, Mohanta K (2012) Use of different microbial probiotics in the diet of rohu, Labeo rohita fingerlings: effects on growth, nutrient digestibility and retention, digestive enzyme activities and intestinal microflora. Aquacult Nutr 18:1–11

Mouriño J, Do Nascimento Vieira F, Jatoba A, Da Silva B, Jesus G, Seiffert W, Martins M (2012) Effect of dietary supplementation of inulin and W. cibaria on haemato-immunological parameters of hybrid surubim (Pseudoplatystoma sp). Aquacult Nutr 18:73–80

Nawaz A, Bakhsh javaid A, Irshad S, Hoseinifar SH, Xiong H (2018) The functionality of prebiotics as immunostimulant: Evidences from trials on terrestrial and aquatic animals. Fish Shellfish Immunol 76:272–278

Newaj-Fyzul A, Adesiyun AA, Mutani A, Ramsubhag A, Brunt J, Austin B (2007) *Bacillus subtilis* AB1 controls *Aeromonas* infection in rainbow trout (*Oncorhynchus mykiss*, Walbaum). J Appl Microbiol 103:1699–1706

Panigrahi A, Kiron V, Satoh S, Hirono I, Kobayshi T, Sugita H, Puangkaew J, Aoki T (2007) Immune modulation and expression of cytokine genes in rainbow trout *Oncorhynchus mykiss* upon probiotic feeding. Dev Comp Immunol 31:372–382

Petersson J, Schreiber O, Hansson GC, Gendler SJ, Velcich A, Lundberg JO, Roos S, Holm L, Phillipson M (2011) Importance and regulation of the colonic mucus barrier in a mouse model of colitis. Am J Physiol Gastrointest Liver Physiol 300:G327–G333

Raulet DH (2004) Interplay of natural killer cells and their receptors with the adaptive immune response. Nat Immunol 5:996–1002

Reyes-Becerril M, Salinas I, Cuesta A, Meseguer J, Tovar-Ramirez D, Ascencio-Valle F, Esteban MA (2008) Oral delivery of live yeast *Debaryomyces hansenii* modulates the main innate immune parameters and the expression of immunerelevant genes in the gilthead seabream (*Sparus aurata* L.). Fish Shellfish Immunol 25:731–739

Ringø E, Song SK (2016) Application of dietary supplements (synbiotics and probiotics in combination with plant products and β-glucans) in aquaculture. Aquacult Nutr 22:4–24

Ringø E, Olsen RE, Mayhew TM, Myklebust R (2003) Electron microscopy of the intestinal microflora of fish. Aquaculture 227:395–415

Ringø E, Sperstad S, Myklebust R, Mayhew TM, Olsen RE (2006) The effect of dietary inulin on aerobic bacteria associated with hindgut of Arctic charr (*Salvelinus alpinus* L.). Aquacult Res 37:891–897

Ringø E, Myklebust R, Mayhew TM, Olsen RE (2007) Bacterial translocation and pathogenesis in the digestive tract of larvae and fry. Aquaculture 268:251–264

Ringø E, Zhou Z, Vecino JLG et al (2016) Effect of dietary components on the gut microbiota of aquatic animals. A never-ending story? Aquacult Nutr 22:219–282

Ringø E, Hoseinifar SH, Ghosh K, Doan HV, Beck BR, Song SK (2018) Lactic acid bacteria in finfish—an update. Front Microbiol 9:1818

Rodriguez-Estrada U, Satoh S, Haga Y, Fushimi H, Sweetman J (2009) Effects of single and combined supplementation of Enterococcus faecalis, mannan oligosaccharide and polyhydroxybutyrate acid on growth performance and immune response of rainbow trout Oncorhynchus mykiss. Aquacult Sci 57:609–617

Rodriguez-Tovar LE, Speare DJ, Markham RJ (2011) Fish microsporidia: immune response, immunomodulation and vaccination. Fish Shellfish Immunol 30:999–1009

Rombout JH, Taverne N, van de Kamp M, Taverne-Thiele AJ (1993) Differences in mucus and serum immunoglobulin of carp (*Cyprinus carpio* L.). Dev Comp Immunol 17:309–317

Safari R, Hoseinifar SH, Kavandi M (2016) Modulation of antioxidant defence and immune response in zebra fish (*Danio rerio*) using dietary sodium propionate. Fish Physiol Biochem. https://doi.org/10.1007/s10695-10016-10253-z

Safari R, Hoseinifar SH, Dadar M, Khalili M (2018) Dietary white bottom mushroom powder improved immunomodulatory and health promoting effects of *Lactobacillus casei* in zebrafish (*Danio rerio*). Int J Med Mushrooms 20:695–704

Sakai DK (1992) Repertoire of complement in immunological defense mechanisms of fish. Annu Rev Fish Dis 2:223–247

Salinas I, Zhang Y-A, Sunyer JO (2011) Mucosal immunoglobulins and B cells of teleost fish. Dev Comp Immunol 35:1346–1365

Sanderson IR, Walker WA (1993) Uptake and transport of macromolecules by the intestine: possible role in clinical disorders (an update). Gastroenterology 104:622–639

Sato A, Mayer WE, Overath P, Klein J (2003) Genes encoding putative natural killer cell C-Type lectin receptors in teleostean fishes. Proc Natl Acad Sci USA 100:7779–7784

Saurabh S, Sahoo PK (2008) Lysozyme: an important defence molecule of fish innate immune system. Aquacult Res 39:223–239

Savan R, Sakai M (2006) Genomics of fish cytokines. Comp Biochem Physiol D 1:9–101

Secombes CJ, Fletcher TC (1992) The role of phagocytes in the protective mechanisms of fish. Annu Rev Fish Dis 2:53–71

Secombes CJ, Zou J, Bird S (2009) Fish cytokines: discovery, activities and potential applications. In: Zaccone G, Meseguer J, Garcia-Ayala A, Kapoor BG (eds) Fish defenses: immunology. Science, Enfield, NH, pp 1–36

Sepahi A, Cordero H, Goldfine H, Esteban MA, Salinas I (2016) Symbiont-derived sphingolipids modulate mucosal homeostasis and B cells in teleost fish. Sci Rep 6:39054

Soltan MA, El-Laithy S (2008) Effect of probiotics and some spices as feed additives on the performance and behaviour of the Nile tilapia, Oreochromis niloticus. Egypt J Aquat Biol Fish 12:63–80

Sveinbjornsson B, Olsen R, Paulsen S (1996) Immunocytochemical localization of lysozyme in intestinal eosinophilic granule cells (EGCs) of Atlantic salmon, *Salmo salar* L. J Fish Dis 19:349–355

Tavares-Dias M, Schalch SHC, Martins ML, Silva ED, Moraes FR, Perecin D (1999) Hematologia de teleósteos brasileiros com infecção parasitaria. I. Variáveis do *Leporinus macrocephalus* Garavello & Britski, 1988 (Anostomidae) e *Piaractus mesopotamicus* Holmberg, 1887 (Characidae). Acta Sci Anim Sci 21:337–342

Tizard IR (2002) Imunologia Veterinária. Uma introdução. Editora Roca, São Paulo. 532 p

Van Doan H, Hoseinifar SH, Tapingkae W, Khamtavee P (2017) The effects of dietary kefir and low molecular weight sodium alginate on serum immune parameters, resistance against *Streptococcus agalactiae* and growth performance in Nile tilapia (*Oreochromis niloticus*). Fish Shellfish Immunol 62:139–146

Wang C, Li J (2015) Pathogenic microorganisms and pancreatic cancer. Gastrointest Tumor 2:41–47

Wershil BK, Furuta GT (2008) 4. Gastrointestinal mucosal immunity. J Allergy Clin Immunol 121: S380–S383

Whyte SK (2007) The innate immune response of finfish—a review of current knowledge. Fish Shellfish Immunol 23:1127–1151

Ye JD, Wang K, Li FD, Sun YZ (2011) Single or combined effects of fructo- and mannan oligosaccharide supplements and Bacillus clausii on the growth, feed utilization, body composition, digestive enzyme activity, innate immune response and lipid metabolism of the Japanese flounder *Paralichthys olivaceus*. Aquacult Nutr 17:e902–e911

Zorriehzahra MJ, Hassan MD, Gholizadeh M, Saidi AA (2010) Study of some hematological and biochemical parameters of Rainbow trout (*Oncorhynchus mykiss*) fry in western part of Mazandaran province, Iran. Iranian J Fish Sci 9:185–198

Feed Additives Impacts on Shellfish Microbiota, Health, and Development

Seyed Hossein Hoseinifar, Maryam Dadar, Hien Van Doan, and Ramasamy Harikrishnan

Abstract Aquaculture industries have received extensive attention and undergone rapid expansion worldwide. Based on FAO statistics, the majority of aquaculture production among different continents takes place in Asian countries, which contribute 90% of worldwide aquaculture production. Intensification of aquaculture practices has often caused diseases to occur, which forces farmers to use chemicals extensively. The overuse or continuous use of antibiotics in aquaculture sanitary management has resulted in the emergence of drug-resistant genes and multiple antibiotic resistance (MAR) bacteria in the aquatic environment of fish and shellfish. During the past decade, numerous studies have considered application of different environmentally friendly feed additives as an alternative to antibiotics in shellfish aquaculture. According to the available literature, functional feed additives including pro-, pre-, and synbiotic are capable of improving digestive function, the utilisation of dietary ingredients, and shellfish performance. Also, numerous reports have shown that functional feed additives can regulate microbial community composition and modulate microbial balance, which will inhibit pathogens, and modulate host

Seyed Hossein Hoseinifar and Ramasamy Harikrishnan contributed equally to the chapter book.

S. H. Hoseinifar (✉)
Faculty of Fisheries and Environmental Sciences, Department of Fisheries, Gorgan University of Agricultural Sciences and Natural Resources, Gorgan, Iran
e-mail: hoseinifar@gau.ac.ir

M. Dadar
Razi Vaccine and Serum Research Institute, Agricultural Research, Education and Extension Organization (AREEO), Karaj, Iran

H. Van Doan
Faculty of Agriculture, Department of Animal and Aquatic Sciences, Chiang Mai University, Chiang Mai, Thailand

R. Harikrishnan
Department of Zoology, Pachaiyappa's College for Men, Kanchipuram, Tamil Nadu, India

N. Derome (ed.), *Microbial Communities in Aquaculture Ecosystems*,
https://doi.org/10.1007/978-3-030-16190-3_7

immune response to exert beneficial effects on aquatic animals. The present chapter of this book summarizes and discusses the available literature regarding possible effects of pro-, pre-, and synbiotics on shellfish growth performance, development, and immune parameters with special focus on mode of action. Also, areas of research that need more attention in future have been highlighted.

1 An Introduction to the Modulation of Health Status in Shellfish: Dietary Approaches

The aquaculture industry has received extensive attention and experienced rapid expansion worldwide (FAO 2016), mainly because of the increasing world population and market demands for healthy seafood on one hand and on the other hand the limitations of wild fish and shellfish populations to meet this increasing demand (Goldburg and Naylor 2005). Among the continents where aquaculture is practiced, Asia (e.g., Bangladesh, China, Thailand, Vietnam) contributes 90% of worldwide production (FAO 2016). Intensification of aquaculture practices has been accompanied by increasing disease prevalence, forcing farmers to use chemicals extensively (Bondad-Reantaso et al. 2005; Hoseinifar et al. 2017a, b, c, d; Nawaz et al. 2018). According to a World Bank report, global shrimp production has had losses of about 3 billion US dollars because of infectious diseases. To prevent occurrence of disease, farmers are asked to maintain water quality, to take care of husbandry as well as to improve immune system performance through environmentally friendly dietary approaches (Elston and Ford 2011; Daniels and Hoseinifar 2014). Excessive and indiscriminate application of chemicals (mainly antibiotics) has been the traditional approach for disease prevention and control in shrimp and prawn aquaculture (Le and Munekage 2004; Mohapatra et al. 2013), and some farmers are using antibiotics as growth promoters (Cabello 2006). The overuse or continuous use of antibiotics in aquaculture practice results in the emergence of drug resistance and multiple antibiotic resistance (MAR) bacteria in fish shellfish in the aquatic environment (Rhodes et al. 2000; Hoseinifar et al. 2017a, b, c, d; Nawaz et al. 2018), which negatively affect the environment as well as terrestrial animals and humans (Cabello 2006). Furthermore, it has been shown that antibiotics administration in fish and shellfish culture alters the gut microbiota, and in turn, exerts negative effects on indigenous human populations (Greenlees 2003; Salyers et al. 2004). In addition, undetected utilisation of antibiotic residuals in food can create large problems of allergy and toxicity (Cabello 2004). Therefore, major fish- and shellfish-producing countries such as Bangladesh, China, Thailand, and Vietnam, as well as importing countries including European Union (EU) countries, the United States, Canada, and Japan, have established strict regulations on application of a wide range of antibiotics and other chemicals such as chloramphenicol, nitrofurans, and parasiticide/fungicide malachite green (Love et al. 2011). Moreover, the incidence of antibiotic residues in culture ponds has been identified as a serious environmental issue. The prophylactic utilisation of antibiotics in commercial aquaculture is a serious concern for food safety because of the persistence of antibiotic residues in flesh (Goldburg et al. 2001).

Therefore, the safety of consumed food requires alternative methods. In many European, American, and Southeast Asian households, shrimp have become common at dinner tables; thus, its safety requires serious consideration (Oosterveer 2006). The use of probiotics as a substitute for antibiotics in aquaculture helps reduce safety concerns (Hoseinifar et al. 2018). To date, there has been no evidence about harmful effects of probiotics incorporated with aquatic products (Modanloo et al. 2017; Ringø et al. 2018; Van Doan et al. 2018). During the past decade, numerous studies regarding the application of different probiotics in shellfish aquaculture have shown a wide range of beneficial effects such as improving soil as well as water quality, influencing host microbiota, contributing to nutrient absorption and enzyme synthesis, enhancing growth rate, inhibiting pathogens, and modulating host immune response (Arndt and Wagner 2007; Merrifield et al. 2010a, b; Hoseinifar et al. 2016; Dawood et al. 2017). Therefore, the aquaculture industry is allowed to use probiotics for bio-control of diseases as a sustainable substitute for antibiotics. Currently, probiotics are commonly used as a functional feed supplement by aquatics feed-producing companies to achieve better growth, improve health, and increase disease resistance (Dawood et al. 2017). It is also well known that probiotics are important in host immune responses (Li et al. 2018a, b; Ringø et al. 2018).

2 Feed Additives, Gut Microbiota, and Growth Performance and Development

According to the available literature, functional feed additives including pro-, pre-, and synbiotics can improve digestive function, utilisation of dietary ingredients, and the performance of aquatic animals (Li et al. 2007; Daniels et al. 2010; Hoseinifar et al. 2017a, b, c, d). Also, the abundant existing literature reveals that probiotics can regulate microbial community composition and modulate microbial balance, which in turn exerts beneficial effects on aquatic animals (Gram et al. 1999; Van Doan et al. 2016, 2017; Dawood et al. 2017; Zheng et al. 2017). A wide range of probiotics including yeasts (*Debaryomyces, Saccharomyces*), and Bacillales as well as lactic acid bacteria (LAB), have been used in shellfish aquaculture (Castex et al. 2014; Daniels and Hoseinifar 2014). In case of shellfish aquaculture, *Bacillus* species and especially *B. subtilis* are perhaps among the most studied probiotics, confirming the positive effects of these species on performance, digestive enzyme activity, and disease resistance in crustaceans (Castex et al. 2014). In a study on the black tiger shrimp (*Penaeus monodon*), Rengpipat et al. (1998) reported that administration of host-associated probiotics (*Bacillus* S11), either live cells or lyophilized, caused significant increase in growth and survival rate when compared with a control group. Also, administration of *Bacillus* sp. in early stages of shrimp development via live feed resulted in notable increase of growth performance and survival rate (Ziaei-Nejad et al. 2006). This result has been attributed to modulation of gut microbiota and enzymatic contribution as well as exclusion of potentially harmful bacteria (Ziaei-Nejad et al. 2006). Numerous probiotics and microbial compounds, including peptidoglycans, lipopolysaccharides, and β-glucans, have also been reported as modulators of shrimp cellular functions (Wang 2007; Bernal et al. 2017; Li et al. 2018a, b). Then, the singular or combined administration of two *Streptomyces* strains, with some genera such as *Lactobacillus* and *Bacillus*, could exert significant effects on the growth parameters, haemocyte count, microflora composition, and immune response of the whiteleg shrimp (*Litopenaeus vannamei*), indicating potential probiotic candidates for improvement of shrimp growth (Das et al. 2010; Bernal et al. 2017).

Furthermore, feed conversion ratio, development, weight gain, and survival rate were significantly higher when compared with control treatment. The best results in case of growth promotion and development were achieved when shrimps were fed with experimental diets supplemented with 1% *Streptomyces* cell mass (strains N7 and RL8) (Das et al. 2010). Besides the aforementioned modes of action, these effects could result from probiotics administrations on the improvement of water quality factors such as decreased concentrations of ammonia and nitrite in pond water compared with control treatment (Silva et al. 2012). Also, the regulation of microflora in the pond water and in the shellfish gut, via maintaining various beneficial bacterial communities and the simultaneous decrease of pathogenic species, are important in improvement of water quality and decreased disruptions by pathogens (Mohapatra et al. 2013). Nonetheless, improvement of water quality is not necessarily the rule for decreasing host mortality, as exemplified with the work of Guo et al. (2009): although the administration of *Bacillus fusiformis* at a 10^5 CFU ml^{-1} concentration in the larviculture system of *Litopenaeus vannamei* significantly increased survival, it had no significant effect on the water quality when compared with control treatment (Guo et al. 2009). Then, several studies showed that exo-enzymes, including lipases proteases, and carbohydrolases could be secreted by *Bacillus* spp. strains that are very effective in breaking down carbohydrates, lipids, and proteins into smaller units (Abraham et al. 1997; Ninawe and Selvin 2009). In addition, these probiotics activities may improve digestion and elevate food absorption, both of which result in the improvement of shrimp growth. Silva et al. (2012) explored the action of probiotics in improving body wet weight gain at first larval phase. Ziaei-Nejad and colleagues (2006) showed that administration of probiotic in the early life stages (nauplii and zoae) of Indian white shrimp (*Fenneropenaeus indicus*) significantly increased weight gain when compared with a group fed with the control diet. The same results were observed in later developmental stages (i.e., mysid and post-larval) (Ziaei-Nejad et al. 2006).

Lactic acid bacteria (LAB)-fed shrimp showed improvement of feed conversion rate (FCR) because *Lactobacillus plantarum* could increase feed utilisation efficiently. In addition, bacterial peptidoglycan (PG) could act as the possible mechanism for LAB-induced improvement in the growth developmental rate. Another study revealed that PG from *Bifidobacterium thermophilum* in supplemented diets of Kuruma shrimp (*Penaeus japonicus*) could significantly improve shrimp survival rates and body weight. Also, *Lactobacillus sporogenes* and *L. acidophilus* significantly increased the growth of *Macrobrachium rosenbergii* post-larvae, although they had no effects on the survival rate (Venkat et al. 2004). Similarly, supplementation of *Fenneropenaeus indicus* larvae diets with *L. plantarum* improved feed utilisation and conversion ratio in comparison with the control treatment.

The combined application of prebiotics and probiotics (synbiotics) is a promising new therapeutic tool in shellfish rearing that has highlighted remarkable improvement of larval survival and growth with the dietary use of functional feed additives compared to nonsupplemented diet. Moreover, greater effects were recorded on growth parameters (e.g., weight/carapace length ratio, moulting condition, weight gain, sustainable growth rate (SGR), FCR, and live carapace length) when prebiotics and probiotics were applied simultaneously in comparison with single administration (Daniels et al. 2010). For example, combined administration of dietary *Bacillus* spp. and mannan oligosaccharides (MOS) showed significant improvement of gut microbiota composition, growth performance, and gut morphology in European lobster (*Homarus gammarus* L.) larvae (Daniels et al. 2010). Comparable successful achievements were reported with the dietary supplementation of *Bacillus* probiotics on black tiger shrimp (*Penaeus monodon*) (Rengpipat et al. 2003) and white shrimp (*Litopenaeus vannamei*) (Li et al. 2007; Wang 2007; Najmi et al. 2018). Also, it has been reported that supplementation diets with the commercial probiotic Previda could induce remarkable compositional shifts of the microbial community in the intestine of shrimp relative to a control group (Anuta et al. 2016). These outcomes are consistent with another study that reported feeding a short-chain fructooligosaccharide (FOS)-supplemented diet which significantly altered the immune parameters as well as the gastrointestinal microbiota of Pacific white shrimp (Li et al. 2007). Considering the significant effects of culture pond water microflora and gastrointestinal microbiota on the growth and development of shrimp, extensive research regarding possible effects of administration of functional feed additives in water or diet is needed. Also, because limited information is available on the indigenous microbial gastrointestinal populations of shrimp, determination of the cooperative function of indigenous gut microbiota on host development is an obvious area for future research (Table 1).

Table 1 Effects of functional feed additives on shellfish growth and development

Kind of feed additive	Inclusion level (CFU/g)	Administration length	Species (g)	Results	References
Bacillus S11	10^{10}	100 days	Black tiger shrimp (*Penaeus monodon*)	↑ Growth, survival rate	Rengpipat et al. (1998)
Commercial *Bacillus*	$7.3 \pm 0.2 \times 10^6$ $1.0 \pm 0.3 \times 10^7$	6 days	Shrimp (*Fenneropenaeus indicus*)	↑ Feed conversion ratio, specific growth rate, and final production	Ziaei-Nejad et al. (2006)
Photosynthetic bacteria and *Bacillus* sp.	10^8 and 10^9	28 days	Shrimp (*Penaeus vannamei*)	↑ Growth performance ↑ Lipase and cellulase activity	Wang (2007)
Bacillus OJ and isomaltooligosaccharides	0, 10^8, and 10^{10}	28 days	Shrimp (*Litopenaeus vannamei*)	↑ Survivals and immune parameters ↓ Counts of total viable bacteria	Li et al. (2009)
Bacillus spp. and mannan oligosaccharides (MOS)	ND	30 days	European lobster (*Homarus gammarus* L.)	↑ Growth performance ↑ Survival and post-larval condition ↓ Reduced *Vibrio* levels	Daniels et al. (2010)
Bacillus subtilis, B. licheniformis, B. pumilus	5.0×10^{10}	ND	White shrimp (*Litopenaeus vannamei*)	↑ Survival and growth of zoea and mysid phase	Silva et al. (2012)
Lactobacillus plantarum	$2–4 \times 10^8$	30 days	*L. vannamei*	↑ Relative growth rate, feed conversion ratio, and survival rates	
Commercial prebiotic *Previdaâ*	0, 0.2, 0.5, 1.0, or 1.6 g kg 1 by weight	35 days	*Litopenaeus vannamei*	→ Survivals and immune parameters → Intestinal microbiota	Anuta et al. (2016)

β-Glucan and microencapsulated probiotics (*Bacillus subtilis* or *Pediococcus acidilactici*)	ND	90 days	*Pacific white shrimp*	↑ Growth performance and body protein, height of intestinal villi ↑ Intestinal lactic acid bacteria ↓ *Vibrio* spp.	
Lactococcus lactis subsp. *lactis*	10^7 or 10^8	56 days	*Litopenaeus vannamei*	↑ Growth performance, intestinal microbiota, digestive enzyme activities, disease resistance	
Streptomyces strains, alone and combined with *Bacillus* and *Lactobacillus*	1×10^9 with 1×10^9	30 days	Shrimp (*Litopenaeus vannamei*)	↑ Growth performance, survivals, and immune parameters	Bernal et al. (2017)
Lactobacillus plantarum, Lactobacillus delbrueckii, Lactobacillus acidophilus, Lactobacillus rhamnosus, Bifidobacterium bifidum, Streptococcus salivarius, Enterococcus faecium	2×10^9	ND	Shrimp (*Litopenaeus vannamei*)	↑ Growth performance, survivals, and immune parameters	Najmi et al. (2018)

3 Pre-, Pro-, and Synbiotics Effects on Immune Response and Disease Resistance in Shellfish

3.1 *Shellfish Immune System*

The shrimp body outer surface, a hard cuticle, is an efficient first line of defense against infectious agents present in the environment (Holmblad and Söderhäll 1999). The digestive tract is also a hostile environment for pathogens as it is paved with chitinous membranes and produces a wide range of acids and proteolytic enzymes (Jiravanichpaisal et al. 2006). Nevertheless, the digestive tract remains a principal entrance for invasive pathogens, so the host organism must have additional mechanisms to repel the pathogens. The immune system generally (1) recognizes foreign materials, (2) transduces a signal to induce production and release of substances that will target foreign materials, and (3) coordinates action to neutralize harmfulness of foreign materials (Holmblad and Söderhäll 1999). Once a pathogen has entered into the host haemocele, the innate immune mechanism or system can start elimination of foreign materials. The host defense mechanism of shellfish is categorized into humoral and cellular components, which are mediated in the haemolymph plasma or executed directly by intact blood cells (Smith and Chisholm 1992). The haemocytes of shellfish are the primary effectors in host defense and are involved in numerous immune processes (Dyrynda et al. 1995). Based on the cytoplasmic granules, the crustacean haemocyte is classified into three types: (1) hyaline, (2) semi-granular, and (3) granular cells (Söderhäll and Smith 1983; Johansson et al. 2000). The shellfish hyaline haemocytes and granular cells have critical functions and are responsible for recognition, participation in apoptosis and phagocytosis, melanisation, encapsulation, nodulation, cytotoxicity, haemolysis, cell adhesion and degranulation, and cell–cell communication (Söderhäll and Smith 1983; Söderhäll and Cerenius 1992; Prapavorarat et al. 2010).

3.2 *Cell-Mediated Immunity*

3.2.1 Immune Cells

Haemocytes such as hyaline, semi-granular, and granular cells are generally considered as functioning in encapsulation and phagocytosis in shrimps (Matozzo and Marin 2010; Hose et al. 1990; Jiravanichpaisal et al. 2006). The latter action has been considered as one of the main cell-mediated immune response in shellfish (Greenberg and Grinstein 2002). The haemocytes are also responsible for hydrolytic and oxidative enzyme activities and the production of superoxide anion (Matozzo and Marin 2010). The shellfish haemocytes have mechanisms to detect and recognize foreign materials that can activate their cellular defense functions and are responsible for the removal of foreign material (Johnson 1987). The haemocytes not only recognize foreign materials but also distinguish between different agents such as an abiotic particle and the potential pathogen. The inflammatory response occurs when the haemocytes migrate to the site of infection and aggregate into haemocytic nodules (Van de Braak et al. 2002) where cell adhesion molecules (e.g., peroxinectin) present and capture pathogens (Jiravanichpaisal et al. 2006).

3.2.2 Humoral Immunity

Pattern Recognition Proteins (PRPs) To initiate an immune response, the host immune system has to recognize the foreign materials entering the body. The shellfish immune cells have membrane-associated pattern recognition proteins (MAPRPs) such as lipopolysaccharide and β-1,3-glucan-binding protein (LGBP), which are of high importance for immune response via binding and recognizing specific compounds in the pathogen cell walls (Roux et al. 2002). Also, it has been suggested that pattern recognition proteins (PRPs) activate the prophenoloxidase (proPO) system (Liu et al. 2009; Vargas-Albores and Yepiz-Plascencia 2000).

Melanisation Melanisation occurs mainly as a result of phenoloxidase (PO) function through hydroxylation of phenols and oxidation of *o*-phenols to quinones. This humoral immune response takes place in response to foreign materials during wound healing (Vargas-Albores and Yepiz-Plascencia 2000). The PO is produced by hydrolysis of the precursor, prophenoloxidase (proPO) (Gollas-Galván et al. 1999). Foreign materials attached onto the naked endocuticle became encapsulated by melanin and are mostly killed during the process (Nyhlén and Unestam 1980).

Prophenoloxidase (proPO) Activating System The proPO and its active form of PO is a melanin-synthesising enzyme with an important role in the immune responses of shellfish. The proPO is present inside shellfish haemocytes, which activate the calcium present in the bacteria and β-glucans, which is required for the conversion of the pro-ppA (Gollas-Galván et al. 1999). The proPO system has a comparatively minor part in the antiviral defense of shrimp (Wang and Zhang 2008). The detailed immune defense mechanisms of proPO in shellfish are currently unknown, but knowledge of these is important for the development of novel health-managing strategies.

3.3 Mechanisms of Action of Probiotics

Studies published on the probiotics mode of action during the past few decades state that probiotics can either affect the host directly or exert the effects through modulation of the culture environment. Probiotics are a key in digestibility and utilisation of feeds, increasing the feed conversion ratio (FCR) and weight gain (WG). The probiotic microbes adhere and colonize the gut of fish and shellfishes, thereby preventing pathogens from inhabiting the gastrointestinal (GI) tract (Montes and Pugh 1993). Probiotics produce several organic acids and hydrogen peroxide, thus reducing the GI pH and inhibiting pathogen proliferation. Moreover, probiotic bacteria are capable of producing natural antibiotics (e.g., bacteriocins) known to inhibit pathogens (Lewus et al. 1991). Probiotics in shellfish could enhance the innate/nonspecific immune status. In addition, probiotics are observed to inhibit viral diseases in shellfish by favouring antiviral effects (Kamei et al. 1988; Direkbusarakom et al. 1998) and to exhibit anticancer effects (Fernandes and Shahani 1990). Also, probiotics can provide better conditions in pond ecosystems through modulation of water microbiota (Boonthai et al. 2011) and improve the water physicochemical parameters (Zokaeifar et al. 2014). Probiotics applied in water reduce its nitrogen and phosphorus concentrations (Daniels and Hoseinifar 2014). Application of probiotics can inhibit the proliferation and growth of pathogens via competitive exclusion (e.g., competition for nutrients and attachment sites), or by modulating the immune system of the host (Chiu et al. 2007; Gullian et al. 2004), which in turn enhances nonspecific immune parameters, mitigates mucosal tissue damage, and improves tissue repair (Eissa and Abou-ElGheit 2014). Overall, adhesion and colonization of probiotics in mucosal surfaces and subsequent competition has been suggested as the main mode of action for pathogen exclusion (Westerdahl et al. 1991) or immune modulation (Salminen et al. 1998).

3.4 Modulation of Gut Microbiota

The intestinal epithelium is a natural barrier of the gastrointestinal (GI) tract, which is the first line of defense against pathogens. The wide range of bacteria colonizing the GI epithelium constitutes the gut microbiota. The gut microbiota has two distinct functions: (1) maintenance of host mucosal immunity (Lazado and Caipang 2014a, b) and (2) providing the host with nutrients and beneficial enzymes (Lazado et al. 2015a, b). More specifically, the gut microbiota contributes to pathogen exclusion (Li et al. 2018a, b), digestive function (Hoseinifar et al. 2017a, b, c, d), epithelial integrity (Ouwehand et al. 1999; Herich and Levkut 2002), and mucosal immunity (Escobar-Briones et al. 2006). Administration of probiotics will modulate gut conditions toward elimination of pathogens (Lee et al. 2000; Vine et al. 2004) by direct and indirect mechanisms: production of antimicrobial compounds including bacteriocins, hydrogen peroxide, lysozymes, proteases, formation of ammonia and diacetyl, and alteration of pH values (Hoseinifar et al. 2018), and by increasing phagocytosis, production of systematic antibodies, as well as local antibodies at mucosal surfaces (Fuller 1992).

3.5 Immunomodulation

Shellfish lack adaptive immunity and mainly depend on innate immune response for resisting pathogens (Young Lee and Söderhäll 2002). Several reports on shrimp revealed that probiotics are capable of modulating the cellular and humoral immune responses (Lazado et al. 2015a, b; Lakshmi et al. 2013; Ninawe and Selvin 2009; van Hai and Fotedar 2010). There are two apparent great interests in shellfish aquaculture during past years: (1) the increased efficiency of probiotic administration in early stages of infection in shrimp highlights its positive effect on the immune system ability to elicit potent responses against pathogens, and (2) this immunomodulation by probiotics is considered as a very promising alternative to antibiotics. The immune cells of shellfish detecting pathogen-associated molecular patterns (PAMPs) exist for both pathogens and probiotic bacteria in such a way that immunomodulation occurs (Lazado et al. 2015a, b). For instance, elevation of reactive oxygen species (ROS) production (Mujeeb Rahiman et al. 2010) and haemocyte count (Rengpipat et al. 2000; Zhang et al. 2011) was observed following treatment with probiotics. Furthermore, available literature confirmed that probiotics administrated to shrimp can increase the prophenoloxidase (*proPO*)-activating system, which is an important innate immune response against microbial infections in invertebrates (Chiu et al. 2007; Gullian et al. 2004), and phagocytosis (Rengpipat et al. 2000; Tseng et al. 2009), as well as upregulation of different immune-related genes (Antony et al. 2011; Dong et al. 2013).

Total haemocyte count (THC) has been reported to increase in *Macrobrachium rosenbergii* following probiotic application (Mujeeb Rahiman et al. 2010). Similar results were obtained in other shrimp species including *Penaeus monodon* (Rengpipat et al. 2000), *P. japonicus* (Zhang et al. 2011), and *Litopenaeus vannamei* (Li et al. 2007). However, there are also some contradictory results revealing that application of *Lactobacillus plantarum* in *Litopenaeus vannamei* culture decreases THC levels (Chiu et al. 2007). In accordance, dietary *Bacillus subtilis* had no notable effect (Tseng et al. 2009). However, dietary administration of *Bacillus* S1 for black tiger shrimp, *Penaeus monodon* (Rengpipat et al. 2000), and *B. subtilis* E20 for the white shrimp, *Litopenaeus vannamei* (Tseng et al. 2009), increased haemocytic phagocytosis. Furthermore, probiotics influence haemocyte respiratory burst (RB) activity (Mujeeb Rahiman et al. 2010).

The prophenoloxidase (proPO) cascade system has developed as a substitute of immunoglobulins (Igs) in the shrimp immune system; which lacks Igs. Therefore, proPO is important for controlling haemolymph bacterial load (Fagutao et al. 2009). It was suggested that feeding on probiotics increased phenoloxidase (PO) activity in *Litopenaeus vannamei* (Chiu et al. 2007; Li et al. 2007; Nimrat et al. 2012, 2013; Tseng et al. 2009; Wang and Gu 2010), *Penaeus monodon* (Rengpipat et al. 2000), *P. japonicus* (Zhang et al. 2011), and *Macrobrachium rosenbergii* (Mujeeb Rahiman et al. 2010). Also, the results of previous studies on different shrimp or prawn species revealed increase of antioxidant enzyme activity (Castex et al. 2009; Gullian et al. 2004; Li et al. 2007; Wang and Gu 2010; Zhang et al. 2011) and catalase (Castex et al. 2009), as well as other immune parameters including lysozyme and nitric oxide

(NO) synthase (Zhang et al. 2011). In addition to improving antioxidant defence, probiotics could reduce the oxidative stress caused by pathogens (Castex et al. 2009). Rengpipat et al. (2000) reported that *Bacillus* sp. improved disease resistance in black tiger shrimp caused by activation of cellular and humoral immune parameters. Similarly, *Lactobacillus plantarum* improved PO, proPO, respiratory burst (RB), and superoxide dismutase (SOD) activities, and clearance efficiency in shrimp against *Vibrio alginolyticus* (Chiu et al. 2007). On the other hand, Zokaeifa et al. reported an increase in the expression of immune-related genes such as proPO, PE, LPS, β-1,3-glucan-binding protein, and serine protein following treatment with *Bacillus subtilis*.

3.6 Probiotics

A wide range of species belonging to the genera *Bacillus* (e.g., *Bacillus subtilis*, *B. licheniformis*, *B. circulans*, *B. coagulans*, *B. clausii*, *B. megaterium*) and *Lactobacillus* have been used in shellfish aquaculture (Daniels and Hoseinifar 2014). Also, the researchers investigated possible effects of different forms of probionts, either live, dead, freeze-dried, or their metabolites (so-called paraprobiotics) on health status and disease resistance of crustaceans (Merrifield et al. 2010a, b). Lactic acid bacteria strains (belonging to genera such as *Streptococcus*, *Leuconostoc*, *Pediococcus*, *Aerococcus*, *Enterococcus*, *Vagococcus*, *Lactobacillus*, and *Carnobacterium*) are known to produce extracellular compounds such as bacteriocins, which have antimicrobial effects (Ringø et al. 2018). Also, as a result of fermentation, they produce short-chain fatty acids, which modulate the host innate immune response (Pandiyan et al. 2013). Ratanapo and Chulavatnatol (1992) reported that feeding black tiger shrimp with probiotics produced a high concentration of lectins, which was efficient against *Vibrio vulnificus*. Therefore, a possible action of this probiotic on the innate immune response is suggested. In collaborative association with pattern-recognition proteins (PRPs), lectins bind to specific carbohydrates present in the pathogen cell wall, then triggering immune response (Marques and Barracco 2000). Also, studies on *Litopenaeus vannamei* and *Penaeus monodon* revealed significant increase of PO enzyme activity following treatment with probiotics (Gullian et al. 2004; Rengpipat et al. 2000). PO catalyses the oxidation of phenolic materials, producing a dark pigment called melanin. Also, PO causes production of antimicrobial agents including quinones and reactive oxygen species (ROS) (Sritunyalucksana and Söderhäll 2000). In two separate studies, Rengpipat et al. (1998a, b, 2000) showed that dietary probiotic notably enhanced innate immune system in *Penaeus monodon* via activation of both cellular and humoral immune parameters. *Penaeus latisulcatus* treated with *Pseudomonas synxantha*- and *P. aeruginosa*-enriched diet reduced hyaline cell (HC) proportion and elevated semi-granular cell (SGC) and THC (Hai et al. 2009). It has been reported that probiotics are capable of increasing total haemocyte count (THC) in shrimp, which per se stimulates the innate immune response against pathogenic infections (Chiu et al. 2007).

3.7 Prebiotics

Prebiotics have been defined as nondigestible carbohydrates that improve health status via the modulation of gut microbiota (FAO 2007). Prebiotics are reported to increase disease resistance, improve nutrient availability, and decrease pathogenic microbiota via alteration of microbial communities toward potentially beneficial populations (Hoseinifar et al. 2015a, b, c, 2017a, b, c, d; Nawaz et al. 2018). Fructooligosaccharides (FOS), galacto-gluco-mannans (GGMs), glucooligosaccharides (GOS), inulin, isomaltooligosaccharides (IMOs), lactose, lactosucrose (LS), mannanoligosaccharides (MOS), oligofructose (OFT), soyabean oligosaccharides (SBOSs), transgalactooligosaccharides (TOS), and xylooligosaccharides (XOS) are nondigestible prebiotics; FOS, GGM, insulin, MOS, and OFT are among the most studied prebiotics (Daniels and Hoseinifar 2014; Ringø et al. 2014). Prebiotics can be used as substrates to promote proliferation and dominance of indigenous probiotics which per se will favour an increase of production of immunostimulatory metabolites such as short-chain fatty acids (Hoseinifar et al. 2017a, b, c, d).

White shrimp fed with dietary FOS exhibited enhanced growth and immune response (Yousefian and Amiri 2009). FOS supplementation diet in *Procambarus clarkii* also increased innate immune response including PO and SOD activities (Dong and Wang 2013). Also, a study on the narrow clawed crayfish, *Astacus leptodactylus*, revealed a modulation of immunity and resistance against *Aeromonas hydrophila* following administration of FOS and MOS (Safari et al. 2014). The FOS-supplemented diet in *Litopenaeus vannamei* improved weight gain and enhanced haemocyte respiratory burst (RB) activity (Li et al. 2007). Also, feeding lobsters with MOS as prebiotics resulted in alteration of the gut microbiota (Daniels et al. 2010).

3.8 Synbiotics

In spite of the benefits reported for probiotics administration in aquaculture, there were some problems such as competition between administrated probiotics and the indigenous microbiota, and consequently, an uncertainty regarding both colonization and dominance in gut environment for the probiotic strain, as well as the possible risk of transfer of antibiotic resistance genes to nonresistant bacteria (Hoseinifar et al. 2015a, b, c). On the other hand, there are some reports regarding instability of colonization of probiotics in gut and significant reduction of its abundance when administration ceased (Rurangwa et al. 2009; Hoseinifar et al. 2015a, b, c). It has been suggested that co-administration of probiotics with suitable prebiotics as substrate will ensure sustainable colonization of the former. Therefore, to overcome the inherent limitations of probiotics, the synbiotic approach (combined administration of probiotics with prebiotics) has been put forward (Ringø and Song

2016). To date, the administration of synbiotics has proven to be more effective as growth promoters and immunostimulants in comparison with single applications of probiotics or prebiotics (Cerezuela et al. 2011), thus highlighting the synergistic effects of synbiotics. However, at present comparatively limited information is available about different aspects of synbiotics administration in aquaculture (Daniels et al. 2010; Zhang et al. 2011; Hoseinifar et al. 2015a, b, c; Rurangwa et al. 2009). It has been suggested that the functionality of synbiotics on the host is directly affected by selected prebiotics as substrate, by dose as well as administration duration (Merrifield et al. 2010a, b; Cerezuela et al. 2011). Also, the duration of feeding with synbiotics can influence the immunomodulatory effects on shrimps (Oktaviana et al. 2014).

Dietary application of a micro-encapsulated synbiotic containing *Bacillus* sp. and OFT resulted in an increased survival, growth performance, THC, PO activity, and disease resistance in *Litopenaeus vannamei* (Munaeni et al. 2014; Zubaidah et al. 2015; Oktaviana et al. 2014). Similarly, administration of another prebiotic (IMO) as substrate for *Bacillus* sp. showed synergistic effects and significantly increased immune parameters as well as disease resistance in white shrimp (Li et al. 2009; Zhang et al. 2012). Also, the administration to *L. vannamei* of synbiotic-containing probiotics SKT-b and oligosaccharide modulated the immune parameters as well as resistance against infectious myonecrosis virus (IMNV) in white shrimp (Septiani 2011). Finally, the application in *L. vannamei* increased biochemical parameters in haemolymph and innate immune response (Ziaeenezhad and Sharifpour 2016).

4 Research Gaps and Future Perspectives

The present chapter highlights the beneficial effects of functional feed additives on the growth, development, and immune response of shellfish. Review of the literature clearly showed that there is limited information available for shellfish as compared with finfish. Nevertheless, the results of previous studies suggest beneficial effects of these feed additives on growth promoters as well as on immunostimulation. Therefore, there should be more research attempts to evaluate different aspects of their administration in shellfish aquaculture. On the other hand, it has been noted that several parameters such as life stage, dosage, administration duration, and the type of additive as well as microbial communities in the gut can affect the impact on host health. Given that, extensive research is required to design optimal formulations of pro-, pre-, or synbiotics and their inclusion in shellfish species health management. Further isolation of host-associated probiotic candidates and determination of their optimum prebiotic partners based on in vitro and in vivo studies (synbiotic) is obviously a promising area of research in shrimp aquaculture. Also, more attention should be given to determining the mode of action of microbial feed additives on the immunity and performance of shellfish.

References

Abraham TJ, Palaniappan R, Dhevendaran K (1997) Epibiotic infestation of luminous bacteria in penaeid shrimp *Penaeus indicus* (H. Milne Edwards) larvae. Oceanogr Lit Rev 3:530

Antony SP, Singh ISB, Jose RM, Kumar PRA, Philip R (2011) Antimicrobial peptide gene expression in tiger shrimp, *Penaeus monodon*, in response to gram positive bacterial probionts and white spot virus challenge. Aquaculture 316:6–12

Anuta JD, Buentello A, Patnaik S, Hume ME, Mustafa A, Gatlin DM III, Lawrence AL (2016) Effects of dietary supplementation of a commercial prebiotic Previda® on survival, growth, immune responses and gut microbiota of Pacific white shrimp, *Litopenaeus vannamei*. Aquac Nutr 22:410–418

Arndt RE, Wagner EJ (2007) Enriched artemia and probiotic diets improve survival of Colorado River cutthroat trout larvae and fry. N Am J Aquac 69:190–196

Bernal MG, Marrero RM, Campa-Córdova ÁI, Mazón-Suástegui JM (2017) Probiotic effect of *Streptomyces* strains alone or in combination with *Bacillus* and *Lactobacillus* in juveniles of the white shrimp *Litopenaeus vannamei*. Aquac Int 25:927–939

Bondad-Reantaso MG, Subasinghe RP, Arthur JR, Ogawa K, Chinabut S, Adlard R, Tan Z, Shariff M (2005) Disease and health management in Asian aquaculture. Vet Parasitol 132:249–272

Boonthai T, Vuthiphandchai V, Nimrat S (2011) Probiotic bacteria effects on growth and bacterial composition of black tiger shrimp (*Penaeus monodon*). Aquac Nutr 17:634–644

Cabello FC (2004) Antibiotics and aquaculture in Chile: implications for human and animal health. Rev Med Chil 132:1001–1006

Cabello FC (2006) Heavy use of prophylactic antibiotics in aquaculture: a growing problem for human and animal health and for the environment. Environ Microbiol 8:1137–1144

Castex M, Lemaire P, Wabete N, Chim L (2009) Effect of dietary probiotic *Pediococcus acidilactici* on antioxidant defences and oxidative stress status of shrimp *Litopenaeus stylirostris*. Aquaculture 294:306–313

Castex M, Daniels C, Chim L (2014) Probiotic applications in crustaceans. In: Merrifield D, Ringø E (eds) Aquaculture nutrition. Wiley, Oxford. https://doi.org/10.1002/9781118897263.ch11

Cerezuela R, Meseguer J, Esteban M (2011) Current knowledge in synbiotic use for fish aquaculture: a review. J Aquac Res Dev 1:1–7. https://doi.org/10.4172/2155-95546.S1-008

Chiu CH, Guu YK, Liu CH, Pan TM, Cheng W (2007) Immune responses and gene expression in white shrimp, *Litopenaeus vannamei*, induced by *Lactobacillus plantarum*. Fish Shellfish Immunol 23:364–367

Daniels C, Hoseinifar SH (2014) Prebiotic applications in shellfish. In: Merrifield D, Ringø E (eds) Aquaculture nutrition. Wiley, Oxford, pp 401–418

Daniels CL, Merrifield DL, Boothroyd DP, Davies SJ, Factor JR, Arnold KE (2010) Effect of dietary *Bacillus* spp. and mannan oligosaccharides (MOS) on European lobster (*Homarus gammarus* L.) larvae growth performance, gut morphology and gut microbiota. Aquaculture 304:49–57

Das S, Ward LR, Burke C (2010) Screening of marine *Streptomyces* spp. for potential use as probiotics in aquaculture. Aquaculture 305:32–41

Dawood MAO, Koshio S, Esteban MÁ (2017) Beneficial roles of feed additives as immunostimulants in aquaculture: a review. Rev Aquac 10:950–974

Direkbusarakom S, Yoshimizu M, Ezura Y, Ruangpan L, Danayadol Y (1998) *Vibrio* spp., the dominant flora in shrimp hatchery against some fish pathogenic viruses. J Mar Biotechnol 6:266–267

Dong C, Wang J (2013) Immunostimulatory effects of dietary fructooligosaccharides on red swamp crayfish, *Procambarus clarkii* (Girard). Aquac Res 44:1416–1424

Dong HB, Su YQ, Mao Y, You XX, Ding SX, Wang J (2013) Dietary supplementation with *Bacillus* can improve the growth and survival of the kuruma shrimp *Marsupenaeus japonicus* in high-temperature environments. Aquac Int 22:607–617

Dyrynda E, Pipe R, Ratcliffe N (1995) Host defence mechanisms in marine invertebrate larvae. Fish Shellfish Immunol 5:569–580

Eissa N, Abou-ElGheit E (2014) Dietary supplementation impacts of potential non-pathogenic isolates on growth performance, hematological parameters and disease resistance in Nile Tilapia (*Oreochromis niloticus*). J Vet Adv 25:712–719

Elston AR, Ford SE (2011) Shellfish diseases and health management. In: Shumway S (ed) Shellfish aquaculture and the environment. Wiley-Blackwell, Ames, pp 359–394

Escobar-Briones L, Olvera-Novoa MA, Puerto-Castillo C (2006) Avances sobre ecología microbiana del tracto digestivo de la tilapia y sus potenciales implicaciones. VIII Simposium internacional de Nutrición Acuícola. 15–17 de Noviembre. Universidad Autónoma de Nuevo León, Monterrey

Fagutao FF, Koyama T, Kaizu A, Saito-Taki T, Kondo H, Aoki T, Hirono I (2009) Increased bacterial load in shrimp hemolymph in the absence of prophenoloxidase. FEBS J 276:5298–5306

FAO (2007) World review of fisheries and aquaculture. The State of World Fisheries and Aquaculture. Food and Agriculture Organization of the United Nations, Rome

FAO (2016) The state of world fisheries and aquaculture 2016. Food and Agriculture Organization of the United Nations, Rome, pp 243–313

Fernandes CF, Shahani KM (1990) Anticarcinogenic and immunological properties of dietary lactobacilli. J Food Prot 53:704–710

Fuller R (1992) History and development of probiotics. In: Fuller R (ed) Probiotics: the scientific basis. Chapman & Hall, London, pp 1–8

Goldburg R, Naylor R (2005) Future seascapes, fishing, and fish farming. Front Ecol Environ 3:21–28

Goldburg R, Elliott MS, Naylor RL (2001) Marine aquaculture in the United States: environmental impacts and policy option. PEW Oceans Commission, Arlington

Gollas-Galván T, Hernández-López J, Vargas-Albores F (1999) Prophenoloxidase from brown shrimp (*Penaeus californiensis*) hemocytes. Comp Biochem Physiol B Biochem Mol Biol 122:77–82

Gram L, Melchiorsen J, Spanggaard B, Huber I, Nielsen TF (1999) Inhibition of *Vibrio anguillarum* by *Pseudomonas fluorescens* AH2, a possible probiotic treatment of fish. Appl Environ Microbiol 65:969–973

Greenberg S, Grinstein S (2002) Phagocytosis and innate immunity. Curr Opin Immunol 14:136–145

Greenlees KJ (2003) Animal drug human food safety toxicology and antimicrobial resistance—the square peg. Int J Toxicol 22:131–134

Gullian M, Thompson F, Rodriguez J (2004) Selection of probiotic bacteria and study of their immunostimulatory effect in *Penaeus vannamei*. Aquaculture 233:1–14

Guo JJ, Liu KF, Cheng SH, Chang CI, Lay JJ, Hsu YO, Yang JY, Chen TI (2009) Selection of probiotic bacteria for use in shrimp larviculture. Aquac Res 40:609–618

Hai NV, Buller N, Fotedar R (2009) Effects of probiotics (*Pseudomonas synxantha* and *Pseudomonas aeruginosa*) on the growth, survival and immune parameters of juvenile western king prawns (*Penaeus latisulcatus* Kishinouye, 1896). Aquac Res 40:590–602

Herich R, Levkut M (2002) Lactic acid bacteria, probiotics and immune system. Vet Med Czech 47:169–180

Holmblad T, Soderhall K (1999) Cell adhesion molecules and antioxidative enzymes in a crustacean; possible role in immunity. Aquaculture 172:111–123

Hose JE, Martin GG, Gerard AS (1990) A decapod haemocyte classification scheme integrating morphology, cytochemistry, and function. Biol Bull 178:33–45

Hoseinifar SH, Esteban MÁ, Cuesta A, Sun Y-Z (2015a) Prebiotics and fish immune response: a review of current knowledge and future perspectives. Rev Fish Sci Aquac 23:315–328

Hoseinifar SH, Mirvaghefi A, Amoozegar MA, Merrifield D, Ringø E (2015b) In vitro selection of a synbiotic and in vivo evaluation on intestinal microbiota, performance and physiological response of rainbow trout (*Oncorhynchus mykiss*) fingerlings. Aquac Nutr 23:111–118

Hoseinifar SH, Mirvaghefi A, Amoozegar MA, Sharifian M, Esteban MÁ (2015c) Modulation of innate immune response, mucosal parameters and disease resistance in rainbow trout (*Oncorhynchus mykiss*) upon synbiotic feeding. Fish Shellfish Immunol 45:27–32

Hoseinifar SH, Ringø E, Shenavar Masouleh A, Esteban MÁ (2016) Probiotic, prebiotic and synbiotic supplements in sturgeon aquaculture: a review. Rev Aquac 8:89–102

Hoseinifar SH, Dadar M, Ringø E (2017a) Modulation of nutrient digestibility and digestive enzyme activities in aquatic animals: the functional feed additives scenario. Aquac Res 48:3987–4000

Hoseinifar SH, Safari R, Dadar M (2017b) Dietary sodium propionate affects mucosal immune parameters, growth and appetite related genes expression: insights from zebrafish model. Gen Comp Endocrinol 243:78–83

Hoseinifar SH, Sun Y, Zhou Z (2017c) Prebiotics and synbiotics. In: Austin B, Newaj-Fyzul A (eds) Diagnosis and control of diseases of fish and shellfish. Wiley, Oxford. https://doi.org/10.1002/9781119152125.ch7

Hoseinifar SH, Sun Y-Z, Caipang CM (2017d) Short chain fatty acids as feed supplements for sustainable aquaculture: an updated view. Aquac Res 48:1380–1391

Hoseinifar SH, Sun Y, Wang A, Zhou Z (2018) Probiotics as means of diseases control in aquaculture: a review of current knowledge and future perspectives. Front Microbiol 9:2429

Jiravanichpaisal P, Lee B, Soderhall K (2006) Cell-mediated immunity in arthropods: hematopoiesis, coagulation, melanization and opsonization. Immunobiology 211:213–236

Johansson MW, Keyser P, Sritunyalucksana K, Söderhäll K (2000) Crustacean hemocytes and haematopoiesis. Aquaculture 191:45–52

Johnson PT (1987) A review of fixed phagocytic and pinocytotic cells of decapod crustaceans, with remarks on haemocytes. Dev Comp Immunol 11:679–704

Kamei Y, Yoshimizu M, Ezura Y, Kimura T (1988) Screening of bacteria with antiviral activity from fresh water salmonid hatcheries. Microbiol Immunol 32:67–73

Lakshmi B, Viswanath B Sai Gopal DVR (2013) Probiotics as antiviral agents in shrimp aquaculture J Pathog: 1–13. Article ID: 424123

Lazado CC, Caipang CMA (2014a) Mucosal immunity and probiotics in fish. Fish Shellfish Immunol 39:78–89

Lazado CC, Caipang CM (2014b) Bacterial viability differentially influences the immunomodulatory capabilities of potential host-derived probiotics in the intestinal epithelial cells of Atlantic cod *Gadus morhua*. J Appl Microbiol 116:990–998

Lazado CC, Caipang CM, Estante EG (2015a) Prospects of host-associated microorganisms in fish and penaeids as probiotics with immunomodulatory functions. Fish Shellfish Immunol 45:2–12

Lazado CC, Caipang CMA, Estante EG (2015b) Host-associated microorganisms of fish and penaeids as probiotics with immunomodulatory functions. Fish Shellfish Immunol 45:2–12

Le TX, Munekage Y (2004) Residues of selected antibiotics in water and mud from shrimp ponds in mangrove areas in Viet Nam. Mar Pollut Bull 49:922–929

Lee YK, Lim WL, Teng AC, Ouwehand EM, Tuomola EM, Salminen S (2000) Quantitative approach in the study of adhesion of lactic acid bacteria to intestinal cells and their competition with enterobacteria. Appl Environ Microbiol 66:3692–3697

Lewus CB, Kaiser A, Montville JJ (1991) Inhibition of food-borne bacterial pathogens by bacteriocins from lactic acid bacteria isolated from meat. Appl Environ Microbiol 57:1683–1688

Li P, Burr GS, Gatlin DM III, Hume ME, Patnaik S, Castille FL, Lawrence AL (2007) Dietary supplementation of short-chain fructooligosaccharides influences gastrointestinal microbiota composition and immunity characteristics of Pacific white shrimp, *Litopenaeus vannamei*, cultured in a recirculating system. J Nutr 137:2763–2768

Li JQ, Tan BP, Mai KS (2009) Dietary probiotic *Bacillus* OJ and isomaltooligosaccharides influence the intestine microbial populations, immune responses and resistance to white spot syndrome virus in shrimp (*Litopenaeus vannamei*). Aquaculture 291:35–40

Li E, Xu C, Wang X, Wang S, Zhao Q, Zhang M, Qin JG, Chen L (2018a) Gut microbiota and its modulation for healthy farming of pacific white shrimp *Litopenaeus vannamei*. Rev Fish Sci Aquac 26:381–399

Li X, Ringø E, Hoseinifar SH, Lauzon H, Birkbeck H, Yang D (2018b) Adherence and colonisation of microorganisms in the fish gastrointestinal tract. Rev Aquac. https://doi.org/10.1111/raq.12248

Liu CH, Chiu CS, Ho PL, Wang SW (2009) Improvement in the growth performance of white shrimp, *Litopenaeus vannamei,* by a protease producing probiotic, *Bacillus subtilis* E20, from natto. J Appl Microbiol 107:1031–1041

Love DC, Rodman S, Neff RA, Nachman KE (2011) Veterinary drug residues in seafood inspected by the European Union, United States, Canada, and Japan from 2000 to 2009. Environ Sci Technol 45:7232–7240

Marques MRF, Barracco MA (2000) Lectins, as non-self-recognition factors, in crustaceans. Aquaculture 191:23–44

Matozzo V, Marin MG (2010) The role of haemocytes from the crab *Carcinus aestuarii* (Crustacea, Decapoda) in immune responses: a first survey. Fish Shellfish Immunol 28:534–541

Merrifield DL, Bradley G, Baker RTM, Davies SJ (2010a) Probiotic applications for rainbow trout (*Oncorhynchus mykiss* Walbaum). Aquac Nutr 16:496–503

Merrifield DL, Dimitroglou A, Foey A, Davies SJ, Baker RTM, Bøgwald J, Castex M, Ringø E (2010b) The current status and future focus of probiotic and prebiotic applications for salmonids. Aquaculture 302:1–18

Modanloo M, Soltanian S, Akhlaghi M, Hoseinifar SH (2017) The effects of single or combined administration of galactooligosaccharide and *Pediococcus acidilactici* on cutaneous mucus immune parameters, humoral immune responses and immune related genes expression in common carp (*Cyprinus carpio*) fingerlings. Fish Shellfish Immunol 70:391–397

Mohapatra S, Chakraborty T, Kumar V, DeBoeck G, Mohanta K (2013) Aquaculture and stress management: a review of probiotic intervention. J Anim Physiol Anim Nutr 97:405–430

Montes AJ, Pugh DG (1993) The use of probiotics in animal food practise. Vet Med 88:282–288

Mujeeb Rahiman KM, Jesmi Y, Thomas AP, Mohamed Hatha AA (2010) Probiotic effect of *Bacillus* NL110 and *Vibrio* NE17 on the survival, growth performance and immune response of *Macrobrachium rosenbergii* (de Man). Aquac Res 41:120–134

Munaeni W, Yuhana M, Widanarni W (2014) Effect of micro-encapsulated synbiotic at different frequencies for luminous vibriosis control in white shrimp (*Litopenaeus vannamei*). Microbiol Indones 8:73–80

Najmi N, Yahyavi M, Haghshenas A (2018) Effect of enriched rotifer (*Brachionus plicatilis*) with probiotic lactobacilli on growth, survival and resistance indicators of western white shrimp (*Litopenaeus vannamei*) larvae. Iran J Fish Sci 17:11–20

Nawaz A, Bakhsh Javaid A, Irshad S, Hoseinifar SH, Xiong H (2018) The functionality of prebiotics as immunostimulant: evidences from trials on terrestrial and aquatic animals. Fish Shellfish Immunol 76:272–278

Nimrat S, Suksawat S, Boonthai T, Vuthiphandchai V (2012) Potential *Bacillus* probiotics enhance bacterial numbers, water quality and growth during early development of white shrimp (*Litopenaeus vannamei*). Vet Microbiol 159:443–450

Nimrat S, Tanutpongpalin P, Sritunyalucksana K, Boonthai T, Vuthiphandchai V (2013) Enhancement of growth performance, digestive enzyme activities and disease resistance in black tiger shrimp (*Penaeus monodon*) postlarvae by potential probiotics. Aquac Int 21:655–666

Ninawe AS, Selvin J (2009) Probiotics in shrimp aquaculture: avenues and challenges probiotics in shrimp aquaculture. Crit Rev Microbiol 35:43–66

Nyhlén L, Unestam T (1980) Wound reactions and *Aphanomyces astaci* growth in crayfish cuticle. J Invertebr Pathol 36:187–197

Oktaviana A, Widanarni, Yuhana M (2014) The use of synbiotic to prevent IMNV and *Vibrio harveyi* coinfection in *Litopenaeus vannamei*. Hayati J Biosci 21:121–126

Oosterveer P (2006) Globalization and sustainable consumption of shrimp: consumers and governance in the global space of flows. Int J Consum Stud 30:465–476

Ouwehand AC, Kirjavainen PV, Shortt C, Salminen S (1999) Probiotics: mechanisms and established effects. Int Dairy J 9:43–52

Pandiyan P, Balaraman D, Thirunavukkarasu R, George EGJ, Subaramaniyan K, Manikkam S, Sadayappan B (2013) Probiotics in aquaculture. Drug Invent Today 5:55–59

Prapavorarat A, Vatanavicharn T, Söderhäll K, Tassanakajon A (2010) A novel viral responsive protein is involved in hemocyte homeostasis in the black tiger shrimp, *Penaeus monodon*. J Biol Chem 285:21467–21477

Ratanapo S, Chulavatnatol M (1992) Monodin-induced agglutination of *Vibrio vulnificus*, a major infective bacterium in black tiger prawn (*Penaeus monodon*). Comp Biochem Physiol B 102:855–859

Rengpipat S, Phianphak W, Piyatiratitivorakul S, Menasveta P (1998a) Effects of a probiotic bacterium on black tiger shrimp *Penaeus monodon* survival and growth. Aquaculture 167:301–313

Rengpipat S, Rukpratanporn S, Piyatiratitivorakul S, Menasveta P (1998b) Probiotics in aquaculture: a case study of probiotics for larvae of the black tiger shrimp *Penaeus monodon*. In: Flegel TW (ed) Advances in shrimp biotechnology. National Center for Genetic Engineering and Biotechnology, Bangkok, pp 177–181

Rengpipat S, Rukpratanporn S, Piyatiratitivorakul S, Menasaveta P (2000) Immunity enhancement in black tiger shrimp (*Penaeus monodon*) by a probiont bacterium (Bacillus S11). Aquaculture 191:271–288

Rengpipat S, Tunyanun A, Fast AW, Piyatiratitivorakul S, Menasveta P (2003) Enhanced growth and resistance to *Vibrio* challenge in pond-reared black tiger shrimp *Penaeus monodon* fed a *Bacillus* probiotic. Dis Aquat Org 55:169–173

Rhodes G, Huys G, Swings J, McGann P, Hiney M, Smith P, Pickup RW (2000) Distribution of oxytetracycline resistance plasmids between aeromonads in hospital and aquaculture environments: implication of Tn1721 in dissemination of the tetracycline resistance determinant Tet A. Appl Environ Microbiol 66:3883–3890

Ringø E, Song SK (2016) Application of dietary supplements (synbiotics and probiotics in combination with plant products and β-glucans) in aquaculture. Aquac Nutr 22:4–24

Ringø E, Dimitroglou A, Hoseinifar SH, Davies SJ (2014) Prebiotics in finfish: an update. In: Merrifield D, Ringø E (eds) Aquaculture nutrition: gut health, probiotics and prebiotics. Wiley, Hoboken, pp 360–400

Ringø E, Hoseinifar SH, Ghosh K, Doan HV, Beck BR, Song SK (2018) Lactic acid bacteria in finfish—an update. Front Microbiol 9:1818

Roux MM, Pain A, Klimpel KR, Dhar AK (2002) The lipopolysaccharide and beta-1,3-glucan binding protein gene is upregulated in white spot virus-infected shrimp (*Penaeus stylirostris*). J Virol 76:7140–7149

Rurangwa E, Laranja JL, Van Houdt R, Delaedt Y, Geraylou Z, Van de Wiele T, Van Loo J, Van Craeyveld V, Courtin CM, Delcour JA (2009) Selected nondigestible carbohydrates and prebiotics support the growth of probiotic fish bacteria mono cultures in vitro. J Appl Microbiol 106:932–940

Safari O, Shahsavani D, Paolucci M, Atash MMS (2014) Single or combined effects of fructo-and mannan oligosaccharide supplements on the growth performance, nutrient digestibility, immune responses and stress resistance of juvenile narrow clawed crayfish, *Astacus leptodactylus leptodactylus* Eschscholtz, 1823. Aquaculture 432:192–203

Salminen S, Bouley C, Boutron-Ruault M, Cumming J, Franck A, Gibson G, Isolauri E, Moreau M, Roberfroid M, Rowland I (1998) Functional food science and gastrointestinal physiology and function. Br J Nutr 80:S147–S171

Salyers AA, Gupta A, Wang Y (2004) Human intestinal bacteria as reservoirs for antibiotic resistance genes. Trends Microbiol 12:412–416

Septiani GR (2011) Pemberian Sinbiotik dengan frekuensi berbeda pada pakan udang vaname *Litopenaeus vannamei* untuk pencegahan IMNV (Infectious Myonecrosis Virus) [Thesis]. Bogor Agricultural University, Bogor

Silva EF, Soares MA, Calazans NF, Vogeley JL, do Valle BC, Soares R, Peixoto S (2012) Effect of probiotic (*Bacillus* spp.) addition during larvae and postlarvae culture of the white shrimp *Litopenaeus vannamei*. Aquac Res 44:13–21

Smith VJ, Chisholm JRS (1992) Non-cellular immunity in crustaceans. Fish Shellfish Immunol 2:1–31

Söderhäll K, Cerenius L (1992) Crustacean immunity. Annu Rev Fish Dis 2:3–23

Söderhäll K, Smith VJ (1983) Separation of the haemocyte populations of and other marine decapods, and prophenoloxidase distribution. Dev Comp Immunol 7:229–239

Sritunyalucksana K, Söderhäll K (2000) The proPO and clotting system in crustaceans. Aquaculture 191:53–69

Tseng DY, Ho PL, Huang SY, Cheng SC, Shiu YL, Chiu CS, Liu CH (2009) Enhancement of immunity and disease resistance in the white shrimp, *Litopenaeus vannamei*, by the probiotic, *Bacillus subtilis* E20. Fish Shellfish Immunol 26:339–344

Van de Braak K, Botterblom MHA, Liu W, Taverne N, Van der Knaap WPW, Rombout JHWM (2002) The role of the haematopoietic tissue in haemocyte production and maturation in the black tiger shrimp (*Penaeus monodon*). Fish Shellfish Immunol 12:253–272

Van Doan H, Hoseinifar SH, Tapingkae W, Tongsiri S, Khamtavee P (2016) Combined administration of low molecular weight sodium alginate boosted immunomodulatory, disease resistance and growth enhancing effects of *Lactobacillus plantarum* in Nile tilapia (*Oreochromis niloticus*). Fish Shellfish Immunol 58:678–685

Van Doan H, Hoseinifar SH, Tapingkae W, Khamtavee P (2017) The effects of dietary kefir and low molecular weight sodium alginate on serum immune parameters, resistance against *Streptococcus agalactiae*, and growth performance in Nile tilapia (*Oreochromis niloticus*). Fish Shellfish Immunol 62:139–146

Van Doan H, Hoseinifar SH, Khanongnuch C, Kanpiengjai A, Unban K, Srichaiyo S (2018) Host-associated probiotics boosted mucosal and serum immunity, disease resistance and growth performance of Nile tilapia (*Oreochromis niloticus*). Aquaculture 491:94–100

van Hai N, Fotedar R (2010) A review of probiotics in shrimp aquaculture. J Appl Aquac 22:251–266

Vargas-Albores F, Yepiz-Plascencia G (2000) Beta glucan binding protein and its role in shrimp immune response. Aquaculture 191:13–21

Venkat HK, Sahu NP, Jain KK (2004) Effect of feeding *Lactobacillus*-based probiotics on the gut microflora, growth and survival of postlarvae of *Macrobrachium rosenbergii* (de Man). Aquac Res 35:501–507

Vine NG, Leukes WD, Kaiser H, Baxter J, Hecht T (2004) Competition for attachment of aquaculture candidate probiotic and pathogenic bacteria on fish intestinal mucus. J Fish Dis 27:319–326

Wang Y-B (2007) Effect of probiotics on growth performance and digestive enzyme activity of the shrimp *Penaeus vannamei*. Aquaculture 269:259–264

Wang Y, Gu Q (2010) Effect of probiotics on white shrimp (*Penaeus vannamei*) growth performance and immune response. Mar Biol Res 6:327–332

Wang W, Zhang X (2008) Comparison of antiviral efficiency of immune responses in shrimp. Fish Shellfish Immunol 25:522–527

Westerdahl A, Olsson J, Kjelleberg S, Conway P (1991) Isolation and characterization of turbot (*Scophthalmus maximus*) associated bacteria with inhibitory effects against *Vibrio anguillarum*. Appl Environ Microbiol 57:2223–2228

Young Lee S, Söderhäll K (2002) Early events in crustacean innate immunity. Fish Shellfish Immunol 12:421–437

Yousefian M, Amiri M (2009) A review of the use of prebiotic in aquaculture for fish and shrimp. Afr J Biotechnol 8:7313–7318

Zhang Q, Tan B, Mai K, Zhang W, Ma H, Ai Q, Wang X, Liufu Z (2011) Dietary administration of *Bacillus* (*B. licheniformis* and *B. subtilis*) and isomaltooligosaccharide influences the intestinal microflora, immunological parameters and resistance against *Vibrio alginolyticus* in shrimp, *Penaeus japonicus* (Decapoda: Penaeidae). Aquac Res 42:943–952

Zhang J, Liu Y, Tian L, Yang H, Liang G, Xu D (2012) Effects of dietary mannan oligosaccharide on growth performance, gut morphology and stress tolerance of juvenile Pacific white shrimp *Litopenaeus vannamei*. Fish Shellfish Immunol 33:1027–1032

Zheng X, Duan Y, Dong H, Zhang J (2017) Effects of dietary *Lactobacillus plantarum* in different treatments on growth performance and immune gene expression of white shrimp *Litopenaeus vannamei* under normal condition and stress of acute low salinity. Fish Shellfish Immunol 62:195–201

Ziaeenezhad S, Sharifpour I (2016) The effects of two synbiotics on biochemical and immunological parameters of white shrimp (*Litopenaeus vannamei*) hemolymph. Iran Sci Fish J 24:169–178

Ziaei-Nejad S, Rezaei MH, Takami GA, Lovett DL, Mirvaghefi A-R, Shakouri M (2006) The effect of *Bacillus* spp. bacteria used as probiotics on digestive enzyme activity, survival and growth in the Indian white shrimp *Fenneropenaeus indicus*. Aquaculture 252:516–524

Zokaeifar H, Babaei N, Saad CR, Kamarudin MS, Sijam K, Balcazar JL (2014) Administration of *Bacillus subtilis* strains in the rearing water enhances the water quality, growth performance, immune response, and resistance against *Vibrio harveyi* infection in juvenile white shrimp, *Litopenaeus vannamei*. Fish Shellfish Immunol 36:68–74

Zubaidah A, Yuhana M, Widanarni (2015) Encapsulated synbiotic dietary supplementation at different dosages to prevent *Vibriosis* in white shrimp, *Litopenaeus vannamei*. Hayati J Biosci 22:163–168

MIX
Papier aus verantwortungsvollen Quellen
Paper from responsible sources
FSC® C105338

If you have any concerns about our products,
you can contact us on
ProductSafety@springernature.com

In case Publisher is established outside the EU,
the EU authorized representative is:
Springer Nature Customer Service Center GmbH
Europaplatz 3, 69115 Heidelberg, Germany

Printed by Libri Plureos GmbH
in Hamburg, Germany